The Dark Side of AI

Wasswa Shafik

The Dark Side of AI

A Human and Societal Perspective

Springer

Wasswa Shafik
School of Digital Science
University Brunei Darussalam
Bandar Seri Begawan
Brunei Darussalam

ISBN 978-3-032-09129-1 ISBN 978-3-032-09130-7 (eBook)
https://doi.org/10.1007/978-3-032-09130-7

This Springer imprint is published by the registered company Springer Nature Switzerland AG
The registered company address is: Gewerbestrasse 11, 6330 Cham, Switzerland

This book is dedicated to my family, whose unconditional love, sacrifices, and unwavering belief in my potential have been the foundation of all my achievements. Their tireless support and encouragement throughout this book-writing journey have inspired me to strive for excellence and persevere in the face of every challenge.

I also dedicate this work to my friends, whose patience, understanding, and constant motivation have been my anchor through the highs and lows of this book journey. Your presence gave me strength during the most difficult times, and your faith in me kept me going when I doubted myself.

To my mentors, lecturers, and teachers throughout this journey, thank you for shaping my intellectual path and for your invaluable guidance over the years. Your wisdom and dedication to knowledge have not only informed this book but also profoundly influenced the way I think, work, and aspire.

To all individuals with any form of disability and marginalized groups, the poor people across the globe who fight to earn a living, and who wish to make it in life, but the conditions fail them, and to those who cultivate life with limited recognition but limitless passion, the seed savers and soil stewards the world forgot, but who never forgot the land.

Finally, I would like to thank everyone who contributed to this journey and made this book possible. I dedicate this book to all aspiring researchers who dare to explore, question, and create. May this book serve as a small contribution to the collective pursuit of knowledge and as an encouragement never to stop learning.

Foreword by Dr. Mueen Uddin

Artificial Intelligence (AI) has emerged as one of the most transformative forces of our era, promising breakthroughs across healthcare, finance, education, and industrial domains. Yet, alongside its benefits, AI also carries profound risks that extend beyond technical vulnerabilities. Bias in algorithms, privacy intrusions, and ethical blind spots often remain hidden until their consequences surface in society. From job displacement and social manipulation to misinformation and threats to democracy, the darker side of AI is increasingly visible and impossible to ignore.

AI algorithms, often developed with the intent to predict, classify, and optimize, can also reinforce inequalities or unintentionally discriminate. When sensitive data are misused, individuals and communities are left vulnerable to exploitation and marginalization. Similarly, AI-driven social manipulation through targeted advertising, political campaigns, or even subtle shifts in recommendation systems has the potential to polarize societies and destabilize democratic structures. These are not hypothetical risks; they are unfolding realities.

From my own perspective as an AI researcher and cybersecurity expert, the darker side of AI lies not only in its misuse but also in our collective complacency. We often celebrate innovation without questioning its long-term societal costs. I believe the greatest danger is not AI itself, but the absence of transparency, accountability, and responsible governance in its deployment. If left unchecked, AI could evolve into a silent weapon shaping opinions, rewriting truths, and eroding human agency, while people remain unaware of the extent of its influence. For me, this is the true dark side of AI: its ability to operate invisibly, yet reshape societies in ways we may only recognize when it is too late.

As we advance, it is imperative to place ethics, responsibility, and human values at the core of AI innovation. This book, *The Dark Side of AI*, takes us on an important journey to examine these challenges and reminds us that while AI can be a powerful tool for progress, it must be guided with wisdom, foresight, and care. Only then can humanity ensure that technology serves us rather than the other way around.

Dr. Mueen Uddin
Associate Professor of Data
and Cybersecurity
College of Computing and IT
University of Doha for Science
and Technology
Doha, Qatar

Foreword by S. Mojtaba Matinkhah

This book, *The Dark Side of AI*, is not a work of alarmist rhetoric. It is the product of an engineer, information technologist, social computer scientist, and AI researcher, Shafik, whose heart and hands have been committed to serving humanity, particularly in Africa, through the responsible use of technology, AI, unmanned aerial vehicles, and digital social platforms. This is the work of an author who has seen the promise of technology firsthand, who has served the needs of communities across Africa, and who now offers this knowledge not to instill fear, but to safeguard hope. After years of work with Non-Governmental Organizations (NGOs) in Uganda and experiences across universities worldwide, he has seen both the hope and the hazards. His message here is grounded not in speculative fear, but in lived experience and deep technical understanding: technology must be handled with prudence, not because it is inherently malevolent, but because its limitations, often hidden to the untrained eye, can translate into significant dangers for human societies when overlooked. This is not a manifesto of resistance to technology; it is a call for prudence in its use, for discernment in its application, and for a vision of development that measures progress not merely in capability but in its alignment with the principles that sustain human dignity.

Here lies a central theme of this work: the technical limits of AI are not trivial; they are foundational. Yet, Shafik recognizes that the "dark side" of AI is not just a matter of philosophical speculation; it is also a matter of hard mathematics, of engineering constraints, and of the logical boundaries that no algorithm can cross. They cannot be patched with faster processors or larger datasets, for they are woven into the very fabric of computation. Beyond bias lies a deeper realm of limitation—Gödel's incompleteness theorems and Turing's undecidability results, which whisper a sobering truth: no formal system, no matter how advanced, can deduce every truth from its axioms, and there are problems that no algorithm can decide. This is not a flaw of current AI; it is a boundary of all possible AI. It means that no machine, however sophisticated, can fully contain morality within a finite set of rules; there will always be moral truths that escape the net of formal reasoning. AI cannot, therefore, be trusted as a self-sufficient moral agent, for it will always operate within a partial view of reality. John Searle's Chinese Room argument reminds us that even if an

AI produces answers indistinguishable from those of a human, it does not mean it understands what it is saying. It manipulates symbols without grasping their meaning. Likewise, the Turing Test, while useful, can be passed by systems that imitate without embodying genuine awareness. This gap between *imitation* and *understanding* is not merely academic; it is a chasm that can mislead societies into believing that machines can replace human judgment in matters of conscience, empathy, and truth. Machine learning models, for example, are trained on human data and therefore inherit human biases. These biases are not always obvious, but they can profoundly affect decision-making, granting loans, recommending parole, prioritizing patients—silently embedding injustice within the machinery of governance and commerce. Without careful design, bias does not just persist; it scales.

Shafik writes not as a prophet of catastrophe, but as a custodian of human well-being. His worldview is rooted in the awareness of the Absolute, in the continuous connection with Him through orthodox religious practices endorsed by authentic traditions. His vision aligns with the beauty of the sovereign Good, because the supreme Principle is not only the Highest but also the most beautiful Essence. *The Dark Side of AI* invites us to face these realities with humility. It urges us to recognize that technological power without metaphysical grounding can lead us into dangerous illusions. The true safeguard is not more code, but deeper wisdom, wisdom drawn from authentic spiritual traditions, where the measure of action is not efficiency alone, but conformity with the Good and the Beautiful. In reading these pages, you will gain not only an understanding of AI's technical shadows but also a compass by which to navigate them, one that points always toward the union of truth, goodness, and beauty.

S. Mojtaba Matinkhah
Yazd University
Yazd, Iran

Foreword by Muchake Brian

Biases in artificial intelligence algorithms represent a principal category of hidden risks and social threats that underlie many other risk factors. Despite being frequently unacknowledged or downplayed, issues, especially those related to system security and personal privacy, draw heavily on ethical considerations. Nonetheless, security and privacy risks can arise directly, in relation to technical vulnerabilities, or indirectly, through misguided AI-assisted attacks. Some forms of market failure, such as widespread job displacement and manipulation of public opinion or consumer demand, likewise have roots in ethical and regulatory shortcomings. Indeed, no risk related to artificial intelligence is purely technical or exclusively derived from technology itself. AI is the field devoted to creating systems that perceive their environment and take actions that maximize their chances of achieving specific goals. Early AI projects concentrated on reasoning and planning; modern AI systems commonly rely on knowledge representation, machine learning, and deep neural networks. A distinction can be drawn among autonomous systems that observe the environment and make decisions; capable systems equipped with advanced reasoning and planning capabilities; and systems with considerable self-awareness and consciousness. Most contemporary "AI systems" encompass only the initial category, autonomous systems.

AI algorithms are designed not only to analyze consumer behaviors and needs but also to classify people and make predictions about their future actions, predictions that may be correct or incorrect, fair or discriminatory. At first glance, it seems advantageous that a model could predict a person's health status, criminal intent, or willingness to repay a bank loan or participate in a political movement. However, sharing such sensitive data with corporations or governments can be dangerous. When an algorithm predicts a person's future criminal intentions, it might unintentionally discriminate against innocent suspects. That classification can also enable the suppression or persecution of specific social groups deemed to diverge from societal norms. Since individuals are often unaware that their data are being sold, fraudulently collected, or even leaked, it is understandable that many people feel overwhelmed and mistrustful of contemporary data dissemination methods. Moreover, this lack of trust is partly due to the adequacy of the laws controlling the collection and use of personal data. In particular, the ethical nature of these services is heavily questioned because many AI methods draw upon big data that might encompass private information about users. This situation can give rise to serious privacy issues connected to the use of individuals' data, the ethical control of such data, and the possibility of granting or refusing AI services based on that data or resulting classifications, especially when predictions turn out to be inaccurate or unfair. Not surprisingly, European data privacy regulation, embodied in the General Data Protection Regulation (GDPR), clearly forbids the use of such sensitive data, among others.

Societal Impact of AI. Many hidden risks associated with AI go far beyond the technical dimension. An often-cited societal effect of the adoption of AI systems is the displacement of workers. The effects of labor displacement have a substantial impact on the masses, especially for those with diminishing skills and digital expertise. People seek different ways to stay relevant at the peril of their physical and mental health. Related to displacement, many jobs or tasks are now at risk of becoming so easy that people become lazy or experience a loss of motivation. Although the development of AI systems can surely generate new job opportunities, they do not compensate for the dramatic effects of displacement. Job paralysis should be considered a real possibility in the future. A subtler aspect of AI's societal impact lies in its potential for social manipulation. A future of hyper-polarization, various conspiracy theories, hyper-partisan politics, and even people being financially invested in a given outcome for their survival becomes a threat to humanity. Humans are hardworking, creative, and resilient; however, the genetics and environmental pressures of our emotions can shape and modulate human decisions and actions. The covert pressure applied by a strong social manipulation utilized by AI systems could modulate our feelings and influence individuals while directing and denationalizing our own decisions. However, the manipulation power of AI can have a more catastrophic impact if deployed in the financial, social, healthcare, political, or generational domains, for example. AI should be carefully scrutinized before being introduced into the market by designing and controlling the regulations associated with it.

Beyond financial aspects, artificial intelligence can also pose risks to freedom and democracy. Numerous studies have demonstrated its potential for behavior manipulation through YouTube recommendations or microtargeted political ads. By fomenting social division and polarization, these techniques erode democratic institutions and can even lead to violence and domestic terrorism. Some experts suggest that such influence operations pose an existential threat to the United States. The intensifying power of AI models such as chat generative pre-trained transformers only compounds these risks. Moreover, AI models can be exploited to generate persuasive misinformation at scale. Malicious actors already use chatbots for phishing and social engineering, and generative style-transfer models can impersonate real individuals. The result can be widespread distrust—not only in media, politicians, or hackers but also in artificial intelligence itself. The underlying challenge is that people typically lack visibility into the decision-making processes of the governments, banks, and technology corporations that shape public discourse. With increasing dependence on AI, the tools for producing and detecting misinformation must advance simultaneously. Otherwise, the fabric of democracies will continue to decay.

Muchake Brian
College of Computing and Information Sciences
Makerere University
Kampala, Uganda

Preface

Due to the advent of highly advanced and sophisticated computer systems that are either designed to simulate characteristics of human intelligence or exhibit some form of problem-solving ability, the quest for artificial intelligence (AI), involving the elaborate study of complex problems offered by these systems, remains an enduring discipline in computer science. In spite of the remarkable research advances and spectacular practical applications of real-world industrial complex problems that AI systems can solve, the study of artificial intelligence raises fundamental issues. These issues span not only the combined sub-discipline of computer science but also philosophy, psychology, linguistics, cognitive neuroscience, and even politics. These complex problems can be loosely classified as issues in theory and issues in application. Primary among issues in theory is knowing what to signify by an "intelligent machine." Is an artificial brain, built using established biological nerve cells, equal to or better than a fully imitated computer system? In addition, researchers in the field of AI and experts in the scientific disciplines in laboratory research on the construction of intelligent machines ignore the association between these fields, but certain discussions revolve around the similarities and the differences they possess, and that people believe they have.

This book does not delve into the technical or algorithmic foundations of artificial intelligence, such as machine learning, generative AI, or large language models, but rather focuses on the broader human, societal, and ethical implications of these technologies. Whether traditional symbolic AI or emerging agentic systems, the consequences for labor, privacy, inequality, and democratic values remain pressing and often overlapping. Critically interrogating the unintended consequences, ethical paradoxes, and societal disruptions brought about by the rapid deployment of AI. While much of the contemporary discourse emphasizes AI's potential for innovation and efficiency, this volume shifts the spotlight to the overlooked or marginalized narratives, those involving algorithmic bias, erosion of privacy, democratic destabilization, and the moral disengagement of automated systems. Given the societal and ethical issues around AI, especially in the workforce, its benefits to diverse other sectors are visible and emerging. Non-experts and many experts have different, sometimes wrong views about AI, often between the "optimists" and "pessimists."

The former group believes AI is beneficial for the world. Through automation and autonomy, if not augmentation, the technology can create more free time for human beings to do what they like to do. AI can also generate economic value and increase the prosperity and happiness of individuals and societies. On the contrary, the latter group argues that AI is dangerous for people, can generate mass unemployment, can be uncontrollable, can be subversive, can destroy individual freedom, or can create super-tasking robots that are simultaneously super-responsible to tackle problems whose essence goes beyond their lines. Either way, they fear that AI will enslave or annihilate humanity for its profit before turning against the Earth and its living beings.

How can we measure the development of AI? Although different kinds of measures are available, none fully represent the complexity of AI and its foreseeable changes in behavior and legal responsibility. Performance on designed tasks is certainly important. It accounts for how much better systems are compared to the status quo of AI, and it is especially interesting when compared to human beings' abilities. Another way to measure AI corresponds to its degree of autonomy, especially in terms of target and adaptive capabilities developed by the system for learning and adaptation. This includes the power of self-reflecting and self-monitoring capabilities. But AI can evolve other traits, particularly those of social intelligence, because it will increase its interactions with human beings. In how many cases of human lives and physical and intellectual activities will AI be employed? The more AI conquers space and importance, potentially at the expense of human lives, the more it will affect presidents and generals, either insidious or devastating, on world power relations. After all, will AI be useful or dangerous?

In recent years, there has been increasing concern about the possibility of creating self-aware or sentient machines. One can intuitively understand an increased earnestness in this inquiry, as machines get more powerful and outperform humans, and as computer science advances to endow things with human-like senses and commonsensical thinking, could there not be a point when machines become more like human beings than mere tools? Hypothetically, it is possible that by self-introspecting, a machine with access to sufficiently rich information could reach a realization about the nature of its self. Current AI programs are typically highly specialized, and their designs usually reflect that specialization. Nonetheless, there is presently no formalized method by which such an AI can evolve into a more general and distributed form, or one that is self-modeling and possibly self-aware. Due to this limitation, AI safety is an easy field to caricature as causing humanity's demise from malevolent machines. The field is not as dystopian as such stereotypes would suggest. However, serious consideration of these problems cannot be avoided; this realization is a driving concern in current AI safety research.

Conceptualizing AI safety concerns necessitates a re-examination of some basic philosophical inquiries; for instance, the basis of reward functions, reasoning about one's knowledge of the world, prediction capabilities, object permanence, and modeling of their very selves. This self-examination should consider the viewpoint of the economy and society in which advanced AI operates. Conversely, the more advanced AI becomes, the more necessary it is for them to consider our societal motivations. Moreover, stating that advanced AI should not be "conscious" is neither an empirically testable hypothesis nor an ethical guide. From an empirical perspective, we do not have a comprehensive theory of consciousness, and there is no agreed-upon way to associate biology with subjective experience empirically. Accordingly, we restrict our discourse exclusively to questions of self-awareness, understanding, and drawing human and societal perspectives. We further describe the behavioral and ethical implications for humans to make progress towards negative and momentous outcomes.

The debate concerning work effects is nuanced and the subject of many studies, both from an economic and technological perspective. An important distinction is a potential change in job numbers; a reduction of employed technology replacing workers through, for instance, more autonomous robots. However, this focus on the number of jobs may be addressing the wrong problem—less work through automation leading to mass leisure implies a question about how we view and value work, and how individuals will want to spend sufficient time away from work. The bigger problem is the increasing income inequality within the workforce. AI alleviates the need for low-skilled work; it cannot replace this work. Although there are no a priori reasons why inequality may increase with AI, there are concerns raised metaphorically by Karl Marx, who foresaw a declining wage share of national income when the final benefits of technological advance are redistributed to a group of capitalists. The mechanism that may lead to increasing inequality mainly arises in the labor market: capital will be able to bid for higher wages while driving down the wages of productive workers. Economically unproductive corporations gain at the expense of productive workers. Inequality could be decreased by strategies such as targeted redistributive policies and by increasing competition within the labor market. In conclusion, the problem is less about losing jobs and mass unemployment, but more about the unequal distribution of income between individuals who actively participate in the labor market.

There are a number of cultural attitudes or types of attitudes that contribute to and reinforce the black box culture. They are exempt from reason or, indeed, any discourse. One of the recognized sacred ideas is that human life has intrinsic value. The idea of regulation is divisive, particularly within the industry, which argues that excessive regulation could significantly hamper AI development and stymie the novel solutions that AI can offer society. Instead, businesses largely argue for a minimalist regulatory approach, enabling a laissez-faire policy in their attempts at AI development. This is particularly important as it is such organizations that are currently leading many of the global AI efforts and that are thereby the recipients of significant venture capital investment. Beyond the business-versus-regulation tension, there are also global-local disparities between nations or even communities regarding how they wish to organize AI research and development. Additionally, it is inappropriate to argue that weaker regulations will promote transnational uneven-playing-field competition between nations where low standards present comparative business advantages when developing and commercializing AI. Furthermore, intergovernmental and international trade negotiations involving discussions of AI principles and regulation are vital both in terms of prevention and establishing shared moral values and norms of cooperation. Establishing global norms for AI compliance mechanisms can aid businesses in coordinating their activities on a global scale, particularly regarding data handling and privacy protection requirements. There are also ways in which governments might intervene, typically involving policies that accelerate the designation of data resources for AI use, encourage the formalization of standardized AI compliance control policies, embark on informational interventions through publicizing industry guidelines, rules, and regulations, and offer support for information-based investments promoting greater transparency, leading toward improved compliance.

The Audience

As artificial intelligence increasingly permeates everyday life, this book explores the shadow it casts across critical human and societal domains, revealing how technologies designed to optimize can also destabilize, marginalize, and dehumanize when left unchecked. This book is written for a wide-ranging audience, including academics, policy-makers, technologists, ethicists, and concerned citizens, who seek to understand how artificial intelligence intersects with fundamental human values, societal structures, and democratic institutions. Whether you are building AI systems or shaping laws that govern them, this volume offers critical insights into the moral and social implications of intelligent technologies. Spanning historical, ethical, political, economic, and regulatory dimensions, it provides a perceptive exploration of AI's darker trajectories. It investigates critical domains such as labor automation, warfare, surveillance, misinformation, and decision-making opacity, highlighting the structural vulnerabilities and power asymmetries that AI can reinforce or exacerbate. Through an interdisciplinary lens, the book synthesizes insights from computer science, philosophy, sociology, law, and public policy, more closely related disciplines. Its scope is global yet inclusive of localized implications, engaging with both high-level frameworks and grounded case studies to map the societal risks and ethical tensions at the heart of today's AI revolution.

The Book Structure

This book is structured into 10 chapters. Chapter 1 establishes the foundational narrative of AI by exploring its origins in logic, philosophy, and computational theory. It presents the historical evolution from rule-based systems to modern neural networks, detailing how ambitions to simulate human intelligence led to the development of autonomous systems. The chapter sets the stage for critical analysis by examining early assumptions, milestones, and turning points that paved the way for both innovation and unforeseen consequences. Chapter 2 introduces AI as a paradoxical force, capable of monumental benefits yet fraught with risks. It lays out the core thesis of the book: that AI, if unregulated or misused, can pose significant threats to human dignity, equity, and democratic institutions. The chapter introduces key concerns, sets expectations for the reader, and outlines the interconnected themes that will be examined in the subsequent chapters.

Chapter 3 investigates the complex moral questions surrounding AI, such as who bears responsibility for algorithmic harm, how to embed values in code, and what constitutes ethical design. It examines ethical frameworks in tension, utilitarianism, deontology, and virtue ethics, as applied to issues like surveillance, bias, and decision automation. Real-world cases are used to highlight how ethics often lags behind technological deployment. Chapter 4 discusses how AI is reshaping employment landscapes, exacerbating income inequality, and creating new forms of precarity. It investigates the displacement of jobs across sectors, the erosion of traditional skill sets, and the concentration of wealth among tech elites. Policy responses such as universal basic income and workforce retraining are evaluated for their potential to address these disruptions.

Chapter 5 reconnoiters the militarization of AI and the ethical quagmires it introduces. Topics include lethal autonomous weapons systems (LAWS), AI-enabled surveillance, and cybersecurity vulnerabilities. It discusses the erosion of accountability in conflict scenarios and the geopolitical arms race to develop smarter, deadlier machines. The chapter warns of the destabilizing potential of AI in the hands of authoritarian regimes and rogue actors. Chapter 6 focuses shift to the manipulation of public opinion through AI-powered technologies such as deepfakes, bots, and algorithmic echo chambers. The chapter investigates how AI is used to distort reality, polarize societies, and undermine democratic processes. It critiques the role of social media platforms and questions the ethical implications of algorithmic amplification of misinformation.

Chapter 7 addresses the opacity inherent in many AI systems, the "Black Box," particularly those using deep learning. It highlights the dangers of unexplainable outcomes in critical areas like healthcare, finance, and criminal justice. The chapter explores the technical and philosophical challenges of achieving transparency, the limits of explainable artificial intelligence (XAI), and the risks of placing blind trust in systems that are not auditable. Chapter 8 surveys international attempts to establish governance frameworks for AI. It compares different regulatory approaches, from the EU's AI Act to ethical guidelines proposed by the United Nations Educational, Scientific, and Cultural Organization (UNESCO), among others. Challenges in enforcement, jurisdiction, and industry compliance are discussed, along with the role of public participation in shaping fair and effective regulations. The chapter emphasizes the urgency of proactive, not reactive, policymaking.

Chapter 9 comes as a transitionary and solution-oriented chapter; it synthesizes previous concerns and offers practical frameworks for responsible AI development. Topics include human-centered design, algorithmic accountability, fairness audits, and participatory design approaches. The chapter promotes interdisciplinary collaboration as essential to developing systems that align with societal values and mitigate harm at scale. Finally, Chap. 10 acts as a prelude to offer a visionary roadmap for AI that serves humanity. It emphasizes the importance of reclaiming agency, ensuring inclusivity, and embedding ethics by design. Drawing on lessons from earlier chapters, it encourages readers, researchers, developers, policymakers, and citizens to engage with AI critically and collaboratively to forge a future that prioritizes justice, well-being, and democratic integrity.

Bandar Seri Begawan, Brunei Darussalam — Wasswa Shafik

Acknowledgments The completion of this book is the outcome of a collective intellectual journey, and I owe a profound debt of gratitude to many individuals and institutions whose contributions, both direct and indirect, have shaped its direction and substance.

I wish to begin by acknowledging Dr. Ali Tufail, Senior Professor Liyanage Chandratilak De Silva, and Dr. Rosyzie Anna Awg Haji Mohd Apong, whose intellectual rigor, methodological clarity, and persistent encouragement provided not only critical guidance but also the academic discipline that underpins this work. His mentorship cultivated the balance between theoretical depth and practical relevance that I sought to achieve throughout this book.

My appreciation also extends to the Series editor and Senior Editor Mayra Castro and publishing professionals who invested their time and expertise in refining the manuscript and ensuring its scholarly relevance. My colleagues, collaborators, and book section reviewers, whose diverse perspectives enriched the conceptual foundations of this project. Also, the academic community at the Dig Connectivity Research Laboratory (DCRLAB) provided a vibrant platform for interdisciplinary discourse at the nexus of artificial intelligence, sustainability, and societal transformation. Their critical reflections, collaborative spirit, and shared pursuit of impactful research profoundly shaped several chapters of this book.

I am grateful to the broader network of scholars and practitioners working at the intersections of technology, sustainability, and governance. Their pioneering work laid a solid foundation upon which this book builds. On a more personal note, I extend heartfelt thanks to my family and close friends, whose patience, resilience, and unwavering encouragement sustained me throughout the long hours of research and writing. Their understanding provided the emotional stability necessary to undertake such a demanding intellectual endeavor.

Finally, I wish to recognize the global community of students, researchers, and policymakers whose engagement with the themes of artificial intelligence, sustainability, and inclusive development continues to push the boundaries of knowledge. This book is written with the hope that it will serve as both a scholarly textbook, reference, and a catalyst for critical debates, informed decisions, and transformative practices in the years ahead. We can make a sustainable difference, no matter how small it is, because it will be of merit to the next generation.

Competing Interests The author has no competing interests to declare that are relevant to the content of this manuscript.

Introduction

Although Artificial Intelligence (AI) is often perceived as a beneficial technology that contributes to the advancement of human society and comforts everyday life, it can also lead to undesirable consequences. The rapid development and adoption of a wide range of AI technologies in various aspects of human life inevitably cause negative side effects that have also become pressing concerns. These concerns arise from the nature of AI technologies and systems as well as their applications. These side effects include the ethical impact on human production and societies, the threat to employment and economic systems, the social and transparent implications, as well as negative influences on mental health. Other considerations include the need for appropriate technical safeguards, products, and demand for proper law and regulation systems. Have you ever thought of such questions, including whether machines can outsmart humans and at what cost? When did you last calculate three or four different numbers without a swift glance at a calculator, yet humans naturally can calculate any given numbers as long as they are not introduced to computing tools? What does it mean for human dignity when invisible, unchallengeable algorithms judge us? Are we truly autonomous when machines curate our thoughts, emotions, and choices? Should a human being's fate be determined by code they cannot question or understand? Is privacy a relic of the pre-AI era, or a right still worth defending? Does it have any substantial effect on our children to use smart devices even if their brains have not fully developed? Can machines ever replace the emotional presence and moral responsibility of a human being? These sampled human-centered questions fight to give your insight into how the majority have given their brain a bed to rest and replace machines to guide their minds, thinking, and emotions. The society-centered structural, institutional, and global-level consequences spark questions on whether AI is enabling a new digital aristocracy, where a few code the future for the many? How can a society function when truth is algorithmically blurred and trust is systemically undermined? Are we witnessing a new form of digital colonialism disguised as innovation? What safety nets or reforms can protect society when the economic ground is shifting beneath millions? Can societies govern technologies they barely understand, and that evolve faster than their laws? These questions will begin to hold some water when we start focusing on using the human and social lens, and addressing these challenges will become more crucial as AI technologies continue to be applied.

Is AI the same as technology? You could be nervous. AI is a branch of information technology that emerged in the mid-20th century, aiming to create computer systems capable of performing tasks typically requiring human intelligence. The initial quest for "strong AI," machines that can reason comparable to or beyond humans, has not been realized. Still, progress has been achieved in developing tools that can support and extend human capabilities across domains based on analysis of data. However, its current progress has worried experts about its abilities. These systems and algorithms underpin important areas of employment and social activity.

The development of AI technologies follows from other areas of information technology: data generation and collection; information and knowledge management; and knowledge automation. The fourth industrial revolution (Industry 4.0) centers on the use of information technology to collect data about manufacturing and control, as well as streaming and managing the data to develop real-time control of manufacturing. Feedback from users is gathered to improve processes and suggest new products and services continually. The development of cyber-physical systems and the Internet of Things (IoT) is an essential component of Industry 4.0. Intelligent systems are integral to this movement. The branch of information technology known as AI has several different subfields. It includes rule-based expert systems—those that use appropriate rules generated by engineers to make decisions. It also includes machine learning systems that learn rules from the data and then use those learnt rules to make predictions about new data. Subsets of machine learning relate to communication (natural language, such as machine translation and chatbots), vision (for example, automatic number plate recognition), and decision-making (such as strategies for playing games such as Go and Chess). Machine learning is also subdivided into supervised learning, unsupervised learning, and reinforcement learning.

The Oxford Dictionary goes further, describing AI as the theory and development of computer systems able to perform tasks normally requiring human intelligence, including visual perception, speech recognition, decision-making, and translation between languages. The term was first articulated in 1956 by John McCarthy during a proposal for a two-month workshop, which he believed could launch a new field of research. The years that followed saw rapid development in AI, marked by periodic oscillations in interest, a phenomenon now termed the AI winter. Expertise generated in the field led to knowledge-based systems with the capacity to solve complex problems not well served by traditional computational methods. The AI brands encountered today vary, general, deep, narrow, super-human, toxic, weak, strong—but essentially embody the same idea: allowing computers to perform tasks associated with human intelligence. At its core, AI seeks to use human intelligence to develop artificial systems capable of performing actions characteristic of humans, or often of superior quality. Hence, it becomes crucial to understand the different types of human intelligence underwriting these functions: social and emotional intelligence, thought intelligence, physical intelligence, and the representation and characterization of thought. Consequently, AI can be distinguished according to the particular facet of thought or intelligence intended for automation. AI technology has made remarkable progress in recent years, with innovations such as automated driving, intelligent self-service, smart cities, and a variety of robots rapidly emerging. Within this development, AI incorporates the research and development of a series of technologies that advance human understanding of intelligence and eventually use computers or machines as carriers to achieve intelligent behavior.

Historically, the idea of creating intelligent machines that can think and reason like humans has been around for centuries. In ancient Greece, the myth of Pygmalion tells of a sculptor who falls in love with a statue he carved, which then comes to life. Similarly, the legend of the Golem describes a clay figure brought to life by magical means. These tales reveal the desire to bring inanimate objects to life. The development of thinking machines began with the invention of the digital computer. British mathematician Alan Turing, widely considered the father of AI, proposed the concept of a universal machine, now known as the Turing machine. In the 1950s, Turing introduced the idea of a test to distinguish between human and machine intelligence, which remains relevant today. The term "AI" itself was coined in 1956 during a workshop at Dartmouth College. AI technologies can be categorized according to their complexity, capabilities, and operational mechanisms. The most recognizable are large language models, used in natural language processing (NLP). These models indeed transform the human experience of text and language by powering chatbots, translations, and explanations, yet their operation may not be immediately apparent due to the complexity of their underlying algorithms. More generally, AI can involve many different types of algorithms, including rule-based, keyword-based, framing-based, clustering, classification, decision trees, learning-to-rank, causal inference, and reinforcement learning models. A broad categorization, however, encompasses all the different applications of AI technologies in an easily understandable way. According to the purpose and application of AI algorithms, technologies, and use cases fall into the categories of general-purpose AI, applied AI, and logic AI. General-purpose AI technologies include facial recognition and advanced game-playing AI, such as DeepMind's AlphaGo. Applied AI refers to the application of general-purpose AI capabilities and human-created knowledge bases for specific purposes; examples are AI in healthcare, law, and customer relationship management (Chatbots). Logic AI is represented by technologies where the AI functions in a closely defined and repetitive manner, such as in sorting emails, making stock trades, or categorizing photos.

AI holds remarkable promise for enabling real progress across a wide range of human endeavors, but it also presents significant risks that must be identified and mitigated to avoid harm. It is essential to acknowledge that beneath much of AI's potential lie hidden algorithmic biases that are seldom disclosed. Institutions like the Defense Advanced Research Projects Agency (DARPA) are actively working on methods to detect and reduce biases in AI systems. However, a complex dilemma arises: as algorithmic equity increases, the accuracy of these systems tends to decline—posing a challenging tradeoff. AI applications confront profound ethical questions, such as determining when access is appropriate or when sensitive data should be exposed to AI systems; for instance, should an AI be allowed to read a patient's health records? Additionally, concerns about pervasive surveillance emerge when AI interprets these records. Autonomous robotic weapon systems highlight the dangers of replacing human judgment with AI decision-making, indicating that moral considerations for AI will continually evolve through successive generations of new technology. Without appropriate governance, AI applications can reinforce unfair power structures, including worker exploitation, income inequality, and social discrimination. Regulatory frameworks that are excessively permissive risk rendering entire groups of workers and consumers vulnerable to data profiling and mass surveillance. To establish global standards for AI regulation, the World Economic Forum proposed the Global Framework for Responsible AI in November 2022. Furthermore, certain future scenarios, such as technological singularity or groups intentionally developing uncontrolled AI models to spread fear, are viewed as dark side risks. Public awareness and education about the consequences and current applications of AI represent effective tools for preventing the algorithmic black hole under discussion by providing adequate knowledge to children, citizens, policymakers, journalists, and society at large.

AI adoption in the workplace continues to generate anxiety and concern. It changes the nature of work, the organization of work, and the employment relationship, and will contribute to increased speed and pressure, a lack of meaning at work, and social isolation. Its effects on jobs generate particular concerns—specifically, the possibility of mass job elimination. Some analysts predict that AI will destroy jobs and create new ones, but that overall, in the mid-term, there will be a growing shortage of qualified people. Others point to the risk that rapid technological changes will knock whole categories of occupations under the intellectual threshold of significant parts of societies. It is estimated that two billion jobs will be lost worldwide in the next 15 years. In addition, the way work is organized is brought under scrutiny, and channels of influence over the employment relationship are disrupted, weakened, or undermined. What happens at work changes and evolves, but it is increasingly influenced by much broader stakeholder groups beyond the employing organization and its immediate sphere of influence. Indeed, many jobs are created and organized by companies classed as taxpayers-of-last-resort, which provide the platform and data needed to deliver such opportunities, control the algorithms that allocate opportunities to individual persons, and determine how the work within a particular category is to be undertaken. The fear of AI and the emerging modes of conducting the employment relationship by the taxpayers-of-last-resort has led to much media fear-mongering, typical of a reaction to technological advances of the past. The conclusion is generally that the vast majority of jobs will be eliminated, as sophisticated robots will take away both manual and intellectual jobs. More cautiously, current developments show that the same fear might have existed when ATMs first emerged, but till now, the cashier has never been fully replaced by a machine. A possible explanation is that ATMs actually made banks more productive, so bank branch managers coped by employing more cashiers.

Social media platforms rely heavily on AI technologies such as personalized content selection, chatbots, digital assistants, translation services, and image recognition. However, more advanced AI applications have also led to detrimental social effects. Social media amplification algorithms can encourage the consumption of misinformation, political extremism, and conspiracy theories, many of which have an anti-scientific or conspiratorial focus. Certain aspects of automated online content moderation prove very controversial. For example, Twitter's image cropping algorithm was shown to have a bias against images of people with darker skin tones. In an effort to profile personalities for marketing, political persuasion, or other purposes, AI systems analyze users' DNA and their social media activity. The increasing pervasiveness of AI technology can also contribute to anxiety, social isolation, and technology addiction. The COVID-19 pandemic accelerated the process of replacing human social interaction with interaction through AI-assisted technology. As social contexts have shifted, the question of AI-mediated human-to-human interactions and their role in human life, including attachment, becomes increasingly important. Further, the use of AI-mediated data collection and surveillance has led to a marked decrease in people's perception and experience of privacy.

AI is recurring among frequently discussed reasons for the growing polarization and division of the population. The explanation is mostly based on the inclusion of AI algorithms in social media. These platforms contribute to the exposure of people to information that fits into their worldview and strengthens it. At the same time, information that contradicts someone's beliefs is diminished or excluded. In addition, in the ultimate consequences of such action, social media platforms influence the division of political parties within the states, as happened in the case of the Brexit referendum in the UK and the US elections. Another significant element affecting the development of conflict is the dissemination of misinformation that is repeated so often that it becomes perceived as factually accurate. Apart from political issues, this is visible during the COVID-19 pandemic in a way that many people became vaccine sceptics or even conspiracy theorists.

The rise of Big Data, through sharing, increased communication, and the development of new technologies, has led to the evolution and emergence of new social behaviors. These changes can be clearly observed in the way of travelling, singing, reading, buying, dating, and many other activities. Not surprisingly, social changes also affect the AI-based optimization of social media sites and search engines. In reality, what began as an innocuous interest in monitoring culture, preferences, microbehaviors, niches of interest, and opinion communities, took on immense proportions when it was discovered that, thanks to the intelligent analysis of the contents posted, it was possible to estimate the behavior and preferences of individuals on the Internet. This, in turn, allowed for the creation of worldwide social sectors that could be manipulated and misguided on a large scale. The idea of being able to influence the behavior of society at large, or at least of the people in the United States during important moments such as presidential elections, is very attractive. The information generated by the Facebook social media platform was used for the US presidential election in 2016, and now, every time elections are held, the public anxiously awaits the announcement of a new scandal. However, the problems go beyond that context. The managed misinformation, careful manipulation of the data reported on social networks, the use of social bots, or the targeted creation of youth or opportunity movements are related problems that are already managing to be incorporated into daily life worldwide.

AI technologies have also impacted human interaction. The use of AI portfolio managers, autonomous vehicles, and in-game virtual managers might reduce the requirement for human interactions. The use of social media, such as Twitter and Facebook, powered by AI technology, is also dramatically affecting human interaction. Adoption of social media has introduced a new driver of conflict and polarization, especially during an election. These regulations are mainly monitored and controlled through AI technologies to detect possible violent events. Such technology works by scanning social media like Twitter, looking for crucial events that can trigger violence. The surveillance camera is connected with an AI technology for monitoring a public place 24 h a day to take unexpected actions to handle the situation. Although the development of AI has resulted in genuine benefits, its excessive use in human life might change human behavior and make them dependent on AI technologies. Such over-dependence might endanger life, as it can lead to loss of human intelligence and knowledge. Misinformation, rumors, and digital manipulation can also be generated through AI. An appropriately crafted message in an election campaign can attract voters, distort information, or promote hatred, violence, or terrorism. Currently, political parties use AI chatbots for their campaigns and make automated phone calls constantly. Nowadays, when COVID-19 has become a huge challenge for the world, technology also plays a crucial role in spreading rumors and misinformation, such as vaccines being unsafe, China's involvement in the Coronavirus, vaccines containing microchips, or the development of COVID-19 vaccines in the USA. On the other hand, the excessive use of technology for entertainment might have negative effects on human psychology, such as technology addiction. Various researchers have demonstrated that technology addiction can lead to anxiety, social isolation, and impairment in social life.

AI may in the future cause at least three mental-health problems: anxiety due to job replacement; social anxiety/introversion due to reduced interpersonal contact; and addiction to AI technologies such as games. The extensive development and implementation of AI technologies have resulted in excessive dependence by humans on the information and communication technologies supported by such an automated system. Although AI has brought great achievements to humanity, the prevalent use of AI has resulted in severe social and psychological problems. Therefore, a study has been conducted to understand the direct and indirect relationship that AI has on humanity and the increasing concern about the manifestation of such social and psychological problems because of the high level of dependence on AI developments and applications. Present-day AI supports all machines, robots, and devices used in the information and communication technology society. Nevertheless, unlike other developed technologies, AI is capable of perceiving its environment, learning from such an environment, and reacting according to perception and learning. Many technological advances are often met with a degree of hesitation or distrust. AI is no exception—a topic often explored in science fiction and closely examined by the media. Such worry is not wholly unfounded, given the ambiguity surrounding its impact on human life, and that a few decades ago, most people did not envision machines taking on tasks that were once solely human. The anxiety has diminished for several technologies, yet AI continues to act as a source of considerable unease.

Severe social isolation is harmful not merely for an individual's well-being but for the well-being of society as a whole. Studies have shown that socially-isolated individuals can have a weaker immune system, experience higher levels of stress, are more vulnerable to developing mental health disorders, are at a greater risk for premature mortality, and face a greater probability of succumbing to cardiovascular diseases and cancer. Further, social isolation—particularly among the young—is related to increases in anti-social behavior, aggression, and crime. When social distancing measures were enforced in the wake of the COVID-19 pandemic, concerns were expressed that technology might have the potential to mitigate the impact of social isolation, for example, by enabling people to remain in contact with friends and family or by supporting the needs of the lonely or housebound. Yet, a large number of commentators have expressed concerns that the way in which social media are designed and provided, together with people's growing dependence on social media and the Internet, has in fact contributed towards increasing loneliness and social isolation within society, particularly among younger people. The power of social media to contribute towards increasing social isolation was evident even prior to the pandemic—Brewis et al. note that the growth of social media has been accompanied by increases in social comparison, a trend that is detrimental to mental health as it can erode self-esteem and cause anxiety and depression.

Interpretations of gambling addiction as a mental health disorder date from the late 1980s. From the perspective of AI's negative human and societal effects, the design of artificial environments may lead to social changes. Today, many activities performed by humans, such as shopping, banking, and communicating, will soon be performed in immersive environments.

Gamers can be at risk of social isolation and a stunted personality. The reasons for this may be the fact that video games are even more intelligent than humans and could replace relationships with other people in their development. Among the emerging social problems related to the use of technology, excessive and addictive video gaming can lead to anxiety or depression. There is concern among experts about the use of new technologies and the risk of dependence. In the field of medicine, this has led to the definition of iDisorder, caused by the addictive use of smart devices or their misuse at the wrong time. A person who develops an addiction to technology is emotionally connected to a particular form of technology. Mental health experts have expressed concerns that the conditioning effect caused by smart technologies may lead to a future of people with limited attention and cognitive abilities. Psychiatrist Manfred Spitzer from the University of Ulm proclaimed that in the near future, people will have the intellect of a dumb fish because they spend so much time with technology. Thomas Goetz has suggested that antisocial tendencies are arising. The ability to interact with people face-to-face at family gatherings or business meetings could be lost. Life away from computer monitors could seem boring and distasteful.

AI is increasingly affecting the whole world, including business, education, investment, healthcare, and government domains. However, many governing bodies are still behind in regulating this evolving technology. The ethical concerns of IP-filtering, bias, and data privacy resemble early internet ethical and political debates; yet, current moves by governments such as the UK, China, Canada, and the USA are seeking to control the evolution of AI by imposing a variety of laws to address these concerns. Various associations, including the Association for Computing Machinery (ACM), have guidelines for the ethical use and development of AI, advocating a culture of peace and tolerance. AI has entered the community's life, impacting social aspects, mental health, government policymaking, and economies. Questions arise about AI's nature and its vulnerabilities. Will AI systems become more capable than humans? Should all services or devices be used with AI? Can AI cause addiction in humans? Will AI alter our nature? Is such a status desirable? Does infiltration of human life by AI create human nature or society anxiety? Will AI eliminate repeated human actions and replace jobs, leading to unemployment? How can AI be accountable if judgment errors occur? Will AI be used in future wars?

Several countries have already enacted AI legislation, with the European Union leading the way. Social and ethical implications have prompted calls for AI regulation in China, the USA, Canada, and other nations. Canada's Directive on Automated Decision-Making establishes guidelines to ensure automation has an appropriate risk level before implementation. The multinational financial services corporation Mastercard has suggested three general areas for future AI law: one regulating the act of building AI—governing how algorithms are programmed and ensuring avoidance of discriminatory behavior; another regulating the use of AI in specific applications, such as facial recognition; and finally, legislation concerning how AI systems are employed—whether to predict the risk of recidivism in criminal cases or automate routine surgery, for example. The Institute of Electrical and Electronics Engineers (IEEE) suggests similar areas of concern and regulation. Despite these proposals, the lack of public awareness about AI contributes to the challenging adoption of such regulatory frameworks.

The Potential Problem of Singularity. Singularity denotes a moment in time when machine intelligence exceeds human intelligence. A super-intelligent machine is deemed capable of recursively improving its capabilities and design. Thereupon, a process of rapid self-improvement initiates, causing the machine to be vastly smarter than humans in a continuum that cannot practically be depicted in a graphical axis of intelligence. The problem of singularity is the hypothetical future challenge that may arise if AI machines ever reach such an extraordinary level of intelligence. In the view of optimistic scholars, a technological singularity may constitute a positive epoch in the history of friendly AI, coinciding with either a runaway increase in human intelligence or a transition to posthuman superintelligence. The sudden arrival of superintelligence would be an intellectual transition comparable in its consequences to the origin of humanity, larger changes to the biosphere, or even the origin of life itself. In contrast, other observers, such as Bill Joy and Stephen Hawking, have stated that the uncontainable self-improving aspect of superintelligence would cause an existential hazard threatening humans, and that it constitutes a greater risk than generally acknowledged by scientists. Regulation and Challenges: AI regulation is the area of endeavor related to the efforts, initiatives, frameworks, guidelines, and even actual laws that regulate or aim to regulate AI. Its aims include attempts to direct the accessibility, use, and development of the emerging technology of AI towards maximizing.

Broadly interpreted, the technological singularity denotes a speculative future point where accelerating technological progress, driven by an intelligence explosion, triggers runaway growth in computation and AI. Proponents argue this would culminate in the creation of hyperintelligences, artificial minds surpassing human cognitive abilities. From the human and societal perspective, singularity is often envisioned as a potential downfall; it poses threats of human extinction or menacing future scenarios. Utopian viewpoints highlight the possibility of zigzagging progress catalyzed by the intelligence explosion and the arrival of superintelligence. The realized side of AI cannot be denied; once AI's capabilities are comparable to humans have been achieved, the journey toward singularity commences. Consequently, researchers have begun to explore the best methods for AI to cohabitate with people that the singularity will affect. In spite of its visionary nature, the notion of singularity carries certain real concerns. Some experts foresee that the era of technological singularity might culminate in what could be termed technological stagnation: the emergence of autonomous entities, robots, or machines equipped with computers, leading to the disappearance of human jobs and engendering profound social transformations. Researchers from prestigious universities worldwide are actively investigating these prospects, cognizant that a nuanced understanding of singularity enhances society's preparedness for the implications of science and technology.

The creation of even the smartest possible AI carries profound implications. AI systems have begun to dictate various aspects of human life, including hiring decisions, political advertisements, defense applications, credit accessibility, and resource distribution. Developing and deploying these technologies requires appropriate cultural, technical, policy, and governance structures. AI has the potential to reshape the distribution of wealth and power among people and nations. Envisioning the trajectory of the human species is therefore crucial: whether as tool-makers, tool-users, tool-lovers, tool-soldiers, or tool-gods, human dignity hinges on deliberately choosing the right future. The development of new AI technologies can take societies toward a future where technology serves those who possess AI-enhanced intelligence, or toward a society in which AI and autonomous systems relieve humans from dangerous, dirty, and dull work, leaving people with jobs that are more difficult but intellectually and socially rewarding. How society utilizes these underlying technologies will determine the outcome.

The emerging power of AI has given rise to concerns among the public about the safety and appropriateness of industry applications. Evidence of the public's perception includes respondents' approval of banning killer robots, restraining AI development, and the fear of job displacement. According to scientific projections, one in three jobs may be replaced by AI technology within the next couple of decades, particularly factory work, driving, and marketing jobs. The actual number of jobs lost to AI depends on various factors, including public perception, market demand, government regulations, policies, responses, and other external conditions. Despite its potential drawbacks, AI also has the power to contribute remarkably to society, especially in medicine (cancer detection, personalized healing) and law enforcement (preventing crime, eliminating jobs with enormous danger levels). However, people still worry whether it is safe to allow AI to take control in specific circumstances. The acceptance of AI deployment varies among industry applications, with the highest being in medicine and the lowest in criminal sentencing. The perceived benefits of AI seem to overshadow the risks or fears AI brings to humanity. Consequently, many researchers advocate for supervised AI technology instead of fully independent and unsupervised frameworks. Moreover, there is a perceived disconnection between the implementation of AI technology and the general public, suggesting that society may not be adequately prepared for the AI wave. Nevertheless, the debate on whether the technological singularity is near continues. Over the product cycle, interest in AI-induced singularity appears to have peaked and even reversed during the COVID-19 epidemic.

As the fantasy surrounding AI is replaced by its instantiations into practical applications, general awareness and knowledge increased, and the majority of people began to interact with AI without identifying it as such. However, the ability to manage AI requires more than awareness and knowledge; it requires an understanding of the real impact of AI on people and society, especially from a negative perspective. The lack of such AI literacy can lead to catastrophic consequences and escalates human–machine interactions to a level that threatens humanity. Owing to its importance, several initiatives have been undertaken to develop specific knowledge associated with AI. In 2019, the World Economic Forum launched the Reskilling Revolution to support the reskilling of more than one billion people affected by job transformation. The European Commission has proposed the European AI Alliance and the Platform on AI to provide a forum for debate and direct support for the development and use of AI in Europe. Furthermore, the European Commission launched the Ethics Guidelines for Trustworthy AI, a document that provides a framework allowing people to evaluate AI systems and understand the risks associated. Similarly, the AI Index Report published by the Stanford Institute for Human-Centered AI addresses the negative implications of AI and highlights how AI perceptions and understandings can also affect its development.

AI (AI) has and is being deployed in countless applications, ranging from emerging and futuristic concepts (e.g., autonomous vehicles, humanoid robots, autonomous weapons, voice assistants in homes, and chatbots assisting on websites) to now basic devices supporting our everyday life (e.g., fraud detection in credit card transactions, social media suggestions and connections, ride-hailing apps, search engine rankings, and yield prediction in agriculture). Given this widespread use of AI by billions of people, it is surprising to see that public understanding of AI is often poor. For example, it is found that many respondents are unaware that AI technology is already in everyday use. Further, participants tend to conflate distinct forms or applications of AI, confuse AI with robotics, and think only of dystopian portrayals. In addition, many have a limited understanding of key terms such as "machine learning." Even though most respondents recognize that AI can have positive and negative impacts, sentiment toward AI tends to be more optimistic. Whereas Americans are more optimistic than pessimistic about the future potential of AI, Asian countries such as India and Singapore are more optimistic than European countries.

Public awareness is an important factor in any debate on AI. An organized effort to help community members understand what types of AI applications exist now and what types do not will help keep society from unmet expectations or unjustified fears. Public awareness and understanding of the human and societal risks of AI will be key to avoiding undesirable outcomes. Community engagement can help broaden the scope of public discussion about AI, ensuring an open discussion of risks at a grassroots level. Such an effort is well-suited to community groups with diverse memberships and a skill in leading and facilitating discussions, such as Rotary International. Three potential benefits emerge:

(i) Engaging diverse backgrounds could reduce the risk of blind spots and groupthink. Discussions should raise questions that might be missed by participants who are non-members of AI societies, lack deep knowledge of AI technologies, or possess a near-term perspective.
(ii) Greater cross-sector engagement may reduce fears or expectations, especially when participants have a better-informed picture of what AI is, what it does, and what it does not.
(iii) Many societal risks require reactions extending beyond the AI field. For example, AI-enabled unemployment may generate demands for economic regulation or societal shifts that go beyond the domain of AI professionals. Hence, broad recognition of the relevant risks, involving community leaders from many sectors, can help ensure that these risks receive the attention they deserve.

An exploration of the potential dark side of AI lends itself to an emphasis on the exposed or latent dangers existing in AI development and deployment. Indeed, the whole of the provided commentary has remained focused on the negative impact of AI on humanity in its broadest sense. The journey has progressed through the phases of definitions, overview, ethical concerns, employment implications, social and mental health effects, regulatory frameworks, future challenges, applications, and public awareness, culminating at the present juncture. Suppose the benefits of any new technology can be maximized while its negative effects are minimized. In that case, the observation can be made, of AI implementation that only in the area of ethics has significant progress been realized. Regulations applicable either specifically to AI or to information technology in general have yet to demonstrate a significant impact on AI's detrimental influence on humanity and society. When public awareness matures to a point enabling full and effective acknowledgment of these harmful effects within the framework of ethical studies, sentiment, and resultant government action, a more balanced evolution and use of AI technology becomes very likely. The content is a must-read for scholars, developers, regulators, and socially conscious readers who want to explore how AI shapes, and sometimes undermines, human dignity, governance, labor, and truth in the digital age.

Bandar Seri Begawan, Brunei Darussalam Wasswa Shafik

Contents

About the Author

Wasswa Shafik born in Makindye (Kampala, Uganda), a Computer Scientist, an Information Technologist, and a Research Director at the Dig Connectivity Research Laboratory (DCRLab), Kampala, Uganda. He received his Bachelor in Information Technology Engineering with a minor in mathematics at Ndejje University, Luweero, Uganda, a Master in information technology engineering (Communication and Computer Networks Option), at Yazd University, Yazd, Islamic Republic of Iran. He pursued his Ph.D. in Digital Science (Computer Science) at the School of Digital Science, Universiti Brunei Darussalam, Brunei Darussalam. His research broadly examines, integrates, and focuses on developing computationally and statistically efficient models and algorithms to address complex questions about artificial intelligence and machine learning problems. His specific research interests include Sustainable Development, Smart Agriculture, Computer Vision, Ecological Informatics, Applied Artificial Intelligence, the Internet of Things, Cybersecurity and Privacy, and Smart Computing. He has authored, co-authored, edited, co-edited, and published hundreds of peer-reviewed books, technical papers, journal papers, and numerous IEEE International Conference papers. His "Hirsch index" is over 20 (*Source* Scopus Author ID 57210203704). He is the co-editor of the "Blockchain Technologies for Smart Circular Economy and Organisational Sustainability" and "Transforming Healthcare Sector Through Artificial Intelligence and Environmental Sustainability" (Ed. Springer Nature, 2025 and 2024), respectively. He is a member of IEEE and has served as a reviewer of several international journals, including Scopus, Compendex (Elsevier Engineering Index), and WoS International Journals. He served in different capacities as department support for Mathematics for Data Science, Physical Computing for Intelligence Systems, Advanced Topics in Computing, Advanced Algorithms, and System Performance and Evaluation. Prior to this, as a department fellow, he served as a researcher associate at the Intelligent Network Laboratory in Iran. He served in different capacities as a Community Data Officer at the Programme for Accessible Health, Communication and Education, Research Associate and Data Manager at Population Services International, Data Manager and Research Assistant at the Socio-Economic Data Center, Research Lead at TechnoServe and Ag. Chief Executive Officer at Asmaah Charity Organisation.

Chapter 1
Awakening the Machine: Tracing the Origins and Evolution of Artificial Intelligence

1.1 Introduction

What if computers could think like people? That idea has been the driving force behind artificial intelligence[1] (AI) research since the dawn of computing. Today's AI technologies transform every corner of our lives, from voice-driven translation to medical diagnostics. Although the concept dates back to 1950, visionary thinkers have been chasing the goal of human-level intelligence for much longer. This narrative traces artificial intelligence's awakening, reveals the conflict inside the intelligence concept, and considers the resulting consequences. The term "artificial intelligence" shorthand the idea that machines can mimic aspects of human thought. Intelligent machines might perform tasks as diverse as planning, speech recognition, or the artful game play that often defines human intelligence. The field absorbed many of its goals and much of its methodological outlook from the let's-make-a-brain approach of the 1940s [1]. Alan Turing famously observed: "Instead of trying to produce a program to simulate the adult mind, why not rather try to produce one which simulates the child's?" The Turing Test, proposed a year later, offers a two-player game to determine whether a computer successfully passes for a human.

Although the concept of AI has existed since antiquity, we consider AI as a field of research since the Turing Test was proposed in 2009. The Dartmouth Summer Research Project in 1956 marked the official beginning of AI as an academic field, combining various topics such as logical reasoning, problem-solving, learning, and advanced natural language understanding. Early research focused on creating symbolic AI, an approach centered on explicit symbolic representations of knowledge, decision rules, and reasoning procedures. BABYLON, developed at the University of Pittsburgh, was an early expert system that could answer natural language medical questions. By the late 1960s and early 1970s, expert systems demonstrated great promise, with new commercial systems emerging rapidly [2]. During the Winter

[1] https://en.wikipedia.org/wiki/artificial_intelligence.

W. Shafik, *The Dark Side of AI*, https://doi.org/10.1007/978-3-032-09130-7_1

of AI, few grants were supported, resulting in minimal advancements, particularly in natural language understanding. Expert systems fell out of favor because their limitations were exposed during real-world trials. The revival of neural networks followed the backpropagation breakthrough, rekindling tremendous interest in AI. The twenty-first century witnessed tremendous growth in AI research, propelled by the combination of big data, enhanced computational power, and algorithmic breakthroughs. The current era is witnessing the rise of Foundation Models, capable of performing numerous tasks without dedicated fine-tuning. Generative AI and foundation models have become dominant forces in the AI landscape during 2022 and 2023 [3].

Yesterday, the idea of machines that are capable of displaying intelligence, such as a human being, was just an idea. Today, we live in a world of AI, and many machines that AI powers are taking care of us in many ways. The notion of machines that can think and behave like a human being has a long-term history, as it is an old sci-fi dream. In fact, humans have been dreaming for two thousand years. Now, the Internet is full of fascinating videos that imagine the shape and events of the exponential rise of AI in the future. AI shows that machines need a Software Brain to be able to perform cognitive tasks. The awakening moment is for the AI machine to wake up from its deep sleep and not just reply to commands and instructions, but to think on its own. The idea of intelligence has existed for many years and has long been an issue of debate over its meaning, as well as over how it is measured and whether it is a unique human characteristic. During different stages of the development of the human race, the thoughts of human beings aimed to understand this idea and investigate it further [4]. Among the things those human beings asked themselves, perhaps the most important question was how to make that creative thing called intelligence in a machine, and this was done by trying to develop a machine that acts like human beings as much as possible.

1.1.1 Early Concepts of Intelligence

The desire for a principled way to think about intelligence predates the modern era, and can be found in works attributed to Descartes, Thurstone, Galton, and others. The study of AI, understood as the engineering of intelligent machines, came into existence as an academic discipline during the 1950s–60 s and was realized in computers that could perform useful tasks requiring intelligence when done by humans. Alan Turing foreshadowed much of this in his 1950 paper 'Computing Machinery and Intelligence,' which introduced "a simple kind of machine, called an 'universal machine,' now known simply as a Turing machine" and started the discussion of whether machines can think. Turing proceeded by defining intelligence operationally, through an imitation game later called the Turing Test. He showed the universal machine's capability to perform any calculation or to carry out any computer program.' Marv Minsky and John McCarthy, founding faculty of MIT's AI Lab, further developed the field through design definitions and academic exploration [5]. The Dartmouth Conference proposal, drafted by McCarthy, Minsky, Nathaniel Rochester, and Claude Shannon, explicitly articulated the hypothesis that every aspect of learning or any other feature of intelligence can be described precisely enough for a machine to simulate.

1.1.2 The Birth of AI: 1950s and 1960s

The inception of artificial intelligence took place in the 1950s and 1960s, when research groups began constructing robots and programs capable of intellectual functioning, thereby realizing Alan Turing's propositions. It entailed a creative surge and institutional organization underpinned by a pioneering scientific vision. AI was conceptualized as a computational endeavor and thus pursued in collaboration with machine-oriented sectors within the cognitive sciences. In 1956, the Dartmouth Conference, organized by John McCarthy and Marvin Minsky, marked the formal establishment of both the term 'artificial intelligence' and a research field capable of supporting large-scale financial and institutional efforts. The conference defined AI as the science of making intelligent machines. In the early 1960s, growing computational power expanded AI applications into computer vision, automatized deductive inference, fruit-picking robots, expert systems, and natural language processing [6]. Commissions investigating the future of science and technology concluded that efforts toward mechanizing higher mental faculties could lead, later in the century, to problem-solving machines capable of inventing, eliciting humor, and recognizing human emotions.

1.2 Key Figures in AI Development

Although the origin and evolution of AI distribution are well-traceable, a look into the world of intelligence before the 1950s–60 s is worthwhile. In The Mathematical Theory of Communication, C. L. Shannon proposed an equivalence of the world as a dictionary. CNA, in turn, proposed the idea that the world is composed of a propositional function named WFF; this WFF can be an antecedent and/or a consequence in all theorems. This implies that the world is currently in a different state than in the previous stage. The propositional function represents an activity that transforms the world from one state to another. These notions associate the concept of intelligence with a function of a different nature from the universe. Alan Turing is considered the father of modern computer science. In 1946, he laid the theoretical foundation by outlining a few simple rules capable of emulating any function of human cognition [7]. In 1950, The Imitation Game was proposed. Formalizing arranged to take up the challenge proposed by John McCarthy, faceted define this label 1 Artificial Intelligence.

John McCarthy is also credited with coining the term Artificial Intelligence. In 1956, together with Frank Rosenblatt, Marvin Minsky, Claude Shannon, and Nathan Rochester, he called a group of scientists and experts from a cross-section of behaviorists, computers, mathematicians, and neurophysiologists to examine Local Computing and intelligent machinery. He proposed Dartmouth College's Artificial Intelligence project, involving a two-month study and teaching conference, the first conference dealing exclusively with AI. In this conference, Mehran Sahami defines the scope and goals of the new discipline: «The scientific understanding of the mechanisms underlying thought and intelligent behavior and their embodiment in machines [8].

This definition means that the field of AI embodies the machinery allowing humans to think logically, analyze images or patterns, build hypotheses from a limited fact set, plan and organize, work in natural language, and solve problems associated with the acquisition of knowledge, data, and skills. Marvin Minsky founded MIT's Artificial Intelligence Laboratory and is universally recognized as a pioneer in the field of Artificial Intelligence. Complementarily to his active involvement in the technological development of AI, he made substantial contributions to the understanding of how the brain works. His tripartite division of brain activity in logical reasoning, problem-solving, and spatial associations set the basis for AI research programs. He studied and developed several models that can perform these functionalities. In parallel, task I-plan machines to execute tasks and achieve goals [9]. This model consisted of the execution of short sequences of actions (the so-called «micro-worlds») until the robot reached its goal.

1.2.1 Alan Turing and the Turing Test

Alan Turing, regarded by many as the father of artificial intelligence, proposed in 1950 that intelligent machines could behave indistinguishably from humans within a few decades. His Turing test concept remains a touchstone for evaluating machine intelligence, yet he questioned the possibility of perfectly indistinguishable machines. Beyond formulating this test, Turing made numerous other contributions to conceptualizing intelligent machines, collectively painting a vision of AI's potential. Although he did not pursue every implication of his ideas, Turing left a blueprint for researchers to explore whether machines could ever think and how such inquiries might advance computer science. In 1959, John McCarthy, who founded machine intelligence research, coined the term artificial intelligence. McCarthy also programmed the first general-purpose AI program and contributed extensively to natural-language processing, design theory, machine learning, robotics, and knowledge representation [10]. Marvin Minsky, co-founder of the AI Laboratory at MIT, was a prolific author and researcher with interests spanning artificial neural networks, embodied learning, and semantic memory.

1.2.2 John McCarthy and the Coining of AI

John McCarthy coined the term artificial intelligence for a 1956 Dartmouth Conference organized with Marvin Minsky, Claude Shannon, and Nathan Rochester. The event is widely accepted as the beginning of AI as a research discipline. The conference proposal explicitly suggested that "every aspect of learning or any other feature of intelligence can in principle be so precisely described that a machine can be made to simulate it." There, participants conjectured that "a significant advance can be made in one or more of these problems if a carefully selected group of scientists work on it together for a summer." The optimism was reminiscent of the cybernetics boom, when machine intelligence was similarly explored. John McCarthy went on to visit Minsky at MIT with the proposal to organize the event that became the Dartmouth Conference [11]. AITP, otherwise known as the Association for the Advancement of Artificial Intelligence (AAAI) and sponsored by the American Psychological Association, was also established at Dartmouth, and then Simon and Newell made presentations there.

1.2.3 Marvin Minsky and the Society of Mind

Marvin Minsky (1927–2016) stands as one of the most profound and comprehensive thinkers in the field of Artificial Intelligence. Many of the ideas and themes that continue to preoccupy AI investigators were foreshadowed in his seminal book The Society of Mind, which appeared in 1988. Turing and McCarthy are usually honoured as the beginning of modern AI, yet Minsky's visionary contribution cannot be underestimated. The theory of Mind as a Society of Agents proposes that the mind is not a singular entity but a complex compilation of numerous interacting agents. It is, in effect, a society of mindless agents. Each of these agents is a mindless, isolated mechanism, but the relationships and interactions among these agents result in the mind's overall functionality. Minsky posits that every kind of behaviour, including thinking, feeling, and sensing, can be implemented through this society of agents [12]. The difference among behaviours lies in the levels of organization and structure of the agents. Consequently, the mind is a society of interacting processes.

1.3 Major Milestones in Artificial Intelligence

Among the most influential milestones in the development of AI was the 1956 Dartmouth Conference, organized by John McCarthy. AI research initially centered on simulating human consciousness and thinking with the brain as a model, but over time, it evolved towards concurrent development of intelligence and consciousness.

In the 1970s, research into expert systems, knowledge bases capable of making intelligent decisions, became a major focus, with the emergence of commercially oriented expert systems spurring widespread AI research. Despite a significant setback during the AI winter, a period of reduced funding and interest, interest in the field has since resumed outside the Anglo-American world. Such landmarks can be defined as the most important events that shaped the field of AI [13, 14]. These include the publication of Alan Turing's 1950 article "Computing Machinery and Intelligence," the Dartmouth Conference in the summer of 1956, John McCarthy's original LISP paper in 1960, Marvin Minsky's 1967 book "Computation: Finite and Infinite Machines," the introduction of the expert system concept in the early 1970s, the AI winter, and IBM Deep Blue's defeat of Garry Kasparov during their rematch in May 1997.

1.3.1 The Dartmouth Conference

The Dartmouth Conference is commonly considered the actual birthplace of artificial intelligence (AI). In 1955, Math Guy drafted a proposal for a research conference. The title of the project was "Artificial Intelligence," and the conference took place the week of June 21, 1956, at Dartmouth College. Math Guy later described the purpose of the workshop as follows: "An attempt will be made to find how to make machines use language, form abstractions and concepts, solve kinds of problems reserved for humans, and improve themselves." Math Guy and Mac Dude, along with other present-day participants, laid out the central questions and problems of modern AI [15]. Since the Dartmouth Conference, numerous seminal events in AI have occurred. No History of AI would be complete without a lengthy discussion of expert systems in the late 1970s, early 1980s, and early 1990s, nor a review of the factors shattering AI expectations during the 1960s and 1970s and more recently. Many expert system successes achieved during the high points of development helped secure the future of the still-maturing AI field, despite the devastating impact of the so-called AI Winter on other parts of it. Yet the History of AI is not yet finished. More milestones await, not all of them in the past [16].

1.3.2 The Rise of Expert Systems

The early 1970s witnessed the rise of expert systems, which were designed to replicate human decision-making in narrow domains. These were the first truly successful commercial applications of artificial intelligence. Expert systems combined a knowledge base, containing domain-specific facts and heuristics, with an inference engine capable of reasoned deductions on that knowledge. The systems had to be provided with knowledge—a substantial task—and were generally limited to relatively narrow domains. One expert system, DENDRAL,[2] was designed to examine mass spectrometer data for organic molecules. Chemists at Heuristic Programming Inc. had been undertaking the laborious task of programming computers to perform data analysis. Because they had no formal training in programming, their efforts were inefficient and often produced code that was difficult for others to understand. The project changed when they met Edward Feigenbaum of Stanford University, who realized the value of the knowledge the Heuristic chemists possessed [16]. Feigenbaum reorganized the effort and provided the team with a sophisticated inference engine that transformed the labor-intensive project into one of the pioneers in expert systems.

1.3.3 AI Winter: Challenges and Setbacks

The Dartmouth Conference launched an era of considerable ambition and optimism. AI in the 1960s and 1970s looked powerful and promising. However, a mixture of expensive failures in the pursuit of grandiose goals, criticism from within and outside the field, and cuts in government funding led to a period of reduced funding and interest. While AI continued to blossom in the 1980s, the limitations of expert systems became apparent. Another so-called "AI winter" ensued in the late 1980s and early 1990s. Growing clamor surrounded the ML approach's rigid statistical and probability models. Skeptics charged historians of AI with cherry-picking groups and timelines, accusing them of downplaying the field's challenges and periods of slow progress. The public's fascination with AI turned a full circle, from awe and wonder to fear and skepticism [17]. Yet the field persisted and quietly evolved. The resurgence of AI in the mid-1990s, with the advent of deep learning, has witnessed machines surpass human champions in strategy and game playing—though this represents just one facet of their burgeoning capabilities.

1.4 Technological Advancements

Nowadays, AI allows machines to act with at least some level of intelligence, exhibiting the human capabilities of reasoning, perception, planning, learning, communicating in natural language, searching knowledge, and problem-solving. Additionally, AI enables machines to detect and classify objects and sounds, computer vision, recognizing and understanding speech, speech recognition and natural language understanding, diagnosing and treating disease, healthcare, planning trips—scheduling, recognizing people by their faces, biometrics, playing games, autonomous driving, and controlling traffic. Artificial intelligence allows machines to act with at least some degree of intelligence. Machines that exhibit human capabilities rely on reasoning, perception, planning, learning, communicating in natural

[2] https://en.wikipedia.org/wiki/dendral.

language, knowledge searching, and problem-solving. Furthermore, AI enables machines to detect and classify objects and sounds (computer vision), recognize and interpret spoken language (speech recognition and natural language understanding), diagnose and treat diseases (healthcare), schedule trips, identify individuals through facial recognition (biometrics), play games, operate autonomously, and manage traffic control [18].

1.4.1 Machine Learning and Neural Networks

The advancement of AI has been anything but slow or limited. From its early beginnings, AI has been awakening, growing in complexity and capability until it is now a common part of our everyday lives, and will soon become an integral element of everything that we do. Machine learning is a subset of AI that focuses on the development of algorithms that enable computers to learn and make predictions or decisions based on data. Machine learning algorithms build a mathematical model based on sample data, known as "training data", in order to make predictions or decisions without being explicitly programmed to perform the task. Neural networks, in particular, form the architectural backbone of many machine learning systems. In essence, a neural network is a group of connected neurons [3]. Hither, every connection, similar to the synapses in a biological brain, can transmit a signal from one neuron to another. The receiving neuron can process the signal and then signal other connected neurons. The "signal" at a connection is a real number, and the output of each neuron is computed by some non-linear function of the sum of its inputs.

1.4.2 Natural Language Processing

One of the key tasks of natural language processing is to transform a natural language description into a machine language description and vice versa. Machines can generate natural language descriptions of their current status or actions through generation, thereby assisting users in human–computer interaction. At the same time, machines can also intelligently understand the natural language texts provided by users through comprehension. When machines comprehend users' natural language texts, they need to extract the desired information for the task from the surface structure described by users and map the meaning to a machine-interpretable representation. At this point, the concept of semantic mapping arises, as it is necessary to transform natural language text into a machine-interpretable representation [19]. This process sounds like the one during awakening, when people use text or speech to provide information about themselves to machines that can understand natural language and process it internally. By repeatedly describing their state to machines, they slowly awaken the machine's memory.

1.4.3 *Computer Vision Breakthroughs*

By the 1980s, the United States was the recipient of a prodigious influx of talent from machine learning and AI, the "foreign legion" mentioned earlier, who continued to reshape the field of computer vision. Their work paved the way for a host of contemporary AI achievements. Breakthroughs encompassed the use of multi-layer feed-forward networks enabled by backpropagation, the Hopkins–Nanjundiah model, Kanade–Lucas–Tomasi tracking, techniques based on Markov random fields, multiple-view geometry, the shape-context descriptor, the dynamic-link architecture, the graph-matching approach, and the seminal test of optical flow detection. More recently, the successful development of deep convolutional neural networks, exemplified by AlexNet, VGG, GoogLeNet, CapsuleNet, ResNet, CondensNet, DenseNet, AgarwoodNet, EfficientNet, Transformer, Vision Transformer, Swin Transformer, ConvNeXt, and CLIP, has propelled deep learning to the forefront of computer vision [20, 21]. Particularly notable is the impact of Vision Transformer, Swin Transformer, and ConvNeXt, which have significantly advanced the field.

1.5 AI in the 21st Century

An examination of twenty-first-century AI reveals both significant achievements and unforeseen consequences. Artificial neural networks, initially proposed in the 1950s, reached a critical depth in the early 2000s. In the public mind, artificial intelligence has indeed "awakened." Machine-learning programs are powering voice recognition devices, intelligent software tools, autonomous vehicles, recommendation engines, and many other applications. Machine intelligence has become part of daily life, and alarm bells are sounding about advanced job automation, algorithmic biases, and lost privacy. There are still many unsolved AI challenges: achieving general intelligence, finding a reliable account of creativity or consciousness, curing machine boredom and suicidal moods, mimicking social and cultural behaviour, and setting and achieving fundamental goals [17]. But the field has moved beyond a belief in some imminent apocalypse, whether of benevolent machines taking care of everything or of infernal machines forever enslaving humanity.

1.5.1 *Deep Learning Revolution*

First coined by Jürgen Schmidhuber in 2015, the phrase Deep Learning Revolution has sparked a new technological race centered on ever-increasing Artificial Intelligence systems. Recent advancements in image understanding, language translation, bioinformatics, and other cognitive tasks have stirred public imagination. The use of deep neural networks in machine learning has been pivotal in powering the current

Artificial Intelligence revolution. The relation between Artificial Intelligence and Machine Learning can be confusing. Artificial Intelligence is the science of building machines that can simulate human intelligence, and Machine Learning, a subfield of AI, is the practice of using data to train models capable of making predictions. Present Time AI includes tasks such as Image Recognition, Speech Recognition and Translation, and Text-to-Everything translation. Future Time AI will involve Text Generation, Video Generation, 3D Model Creation, and Image Inpainting [22]. This progression toward more human-like tasks explains the buzz surrounding Recent Advances in Deep Learning, more specifically, Transformer models such as BERT,[3] GPT-3, and the Imagen series from Google.

1.5.2 AI in Everyday Life

AI increasingly influences modern life. Smartphone assistants perform tasks through natural conversation. Home cameras detect unidentified faces and every movement of residents and visitors. Automatic translators enable real-time communication, intelligible even during diplomatic visits. ATM and credit card purchases require only face validation. Self-driving cars alert to the presence of pedestrians and other vehicles, and some are already operating on the roads in several countries. AI is also essential to a variety of other fields. It assists physicians and medical personnel in diagnosis and treatment, aids in production planning and stock control, generates advertising campaigns, guards' oil and gas production plants, and automates numerous service sector operations [2].

1.5.3 Ethical Considerations in AI

There are significant and valid concerns about whether the power of AI may be channeled morally and ethically that engenders the benefit rather than destruction of the natural and social world. The fear surrounding AI's potential to harm—whether through malevolence or neglect—is illustrated by the repeated portrayal of sentient machines as apocalyptic villains. These negative representations are overwhelmingly dominant, and they can have a palpable, long-term effect on public understanding and support for AI. A dystopian view of AI, combined with ignorance about current AI capabilities, naturally can lead to abject fear of technologies that might not ever be created. The disparity between science fiction AI and contemporary AI is pronounced and can hardly be dismissed as unimportant. Surveys demonstrate that the public is far more supportive of AI technologies when they are perceived as augmenting and assisting humans rather than replacing and competing with them [23, 24]. In addition to fear, an overly hopeful attitude also can distort AI-related investments, research,

---

[3] https://en.wikipedia.org/wiki/BERT_(language_model).

and even social acceptance. How the media and fiction portray AI matters to how society prepares for its influence.

1.6 Understanding Generative Models

Even though there are many different methods used in machine learning, they all try to do similar things. At the core, generative AI has two tasks. The first is to grasp the essence of a user prompt. The second is to use this understanding to create a logical, novel output. Getting these two right determines how successfully artificial intelligence can do the job it was built for. Unlike traditional information systems, which handle the two tasks at once, generative AI breaks down the user prompt to truly get what the user wants. With this knowledge, it then generates the answer in a new and creative way. Because generative AI needs to perform these two tasks one after the other, the methods vary. For instance, a text-summarizing tool simplifies content without really understanding it, since it draws from the user prompt itself. Traditional information systems usually solve their problems sequentially, and generative AI also achieves this by first understanding the text and then generating a summary. The two tasks in generative AI, digesting the user prompt and making an appropriate output, can get as complex as one wants. A simple step to grasp the prompt might involve just translating the language to English first, then dealing with emotions, topics, or specific events. More layers, like explaining or dubbing the user's feelings or the text's topic, can be added [25]. The variety of techniques for these tasks is wide. For translation from any language to English, technologies like Recurrent Neural Networks (RNNs), Long Short-Term Memory (LSTM) networks, Gated Recurrent Units (GRUs), and transformers in a sequence-to-sequence setup can be used.

1.6.1 Types of Generative Models

Deep learning models have evolved to be able to generate new content or data. Several types of models that do just that can be found. The following explores the most notable ones: generative adversarial networks (GANs), variational autoencoders (VAEs), and transformer-based models. GANs usually consist of two models, a generator and a discriminator, that work opposite to one another and compete. The generator creates new data, and the discriminator classifies it as either real or fake. VAEs somehow resemble a classical encoder–decoder architecture. In a first step, input data is encoded into a latent space, usually of lower dimension but possibly of higher dimension. Second, points in latent space are sampled and decoded back into the new data space. VAEs aim to generate the fullest representation possible of the training data by maximizing the log-likelihood of the training data given the parameters of the model [26]. With transformer models, the goal is to predict the next item in a sequence given the preceding items: by maximizing the probability of the correct label, transformers can be trained to generate sequences, one element at a time, iteratively.

1.6.2 Applications of Generative AI

Generative AI applications offer solutions to various issues faced by society. This wide array of use cases is reflected in the large number of companies working in the generative AI domain, many of which provide tools that the general population can use without technical knowledge. This includes well-known generative transformers such as ChatGPT, Google Bard, and Microsoft's Bing Chat for text; DALL·E, Stable Diffusion, and Midjourney for images; and AIVA and Jukebox for music composition. Startups also serve specific areas and niches, such as Reface for face-swapping videos, Bling for removing backgrounds from pictures, Looka for logos, Contlo for generating marketing email text, and Productive for writing code. Generative models are as versatile as human imagination [27].

1.7 Human Interaction with Artificial Intelligence

The empirical view emphasizes the role of the user in the generation of content. Concepts like usability, successful interaction, and enjoyable human–computer synergy are very important. The main questions focus on collaborative co-creation, i.e., how humans engage with the model. Integration is a decisive factor in practical use. Users also express a need for a more seamless and streamlined experience for generating AI-generated content. The aim is to develop more natural and intuitive text-to-image tools that enable easy generation of specific handles, such as human faces and facial expressions. Controllability is another highly desired feature of generative workflows.

1.7.1 User Experience and Design

Generative AI has garnered significant attention since the public release of ChatGPT in November 2022. Initially perceived as a side effect of deep learning improvements, generative AI has become a global phenomenon, into the universe of viruses and memes. What makes generative AI so fascinating and powerful is the potential to use it in creative processes with the help of neutrality, objectivity, and superiority. It can also be tissue paper for the human approach in practical routines and become a black hole for human intelligence, critical thinking, and moral values. Generative AI can be both a blessing and a curse; depending on the future procedures and methods, it can be a human-friendly tool or a human-hostile agent [28]. The question remains:

what about the users and human factors? How can generative AI impact the user side of the system and also the human side of society? The two fundamental questions underlying the design of generative AI systems pertain to the target users and the type of content these models should generate for their users, as summarized in Table 1.1.

The term "generative AI" broadly refers to the class of algorithms capable of generating new content, new images, text, music, or even code, when provided with context or an input query. This capability stems primarily from large-scale training: the models learn to recognize data patterns by processing vast quantities of online content. Once trained, they can then generate brand new content aligned with the input query and also fall within the class of all the patterns they have learned. These capabilities enable the creation of AI-powered tools such as chatbots and text-to-image generators. However, such extensive training has attracted significant criticism

Table 1.1 Human-centric impacts of generative AI across key social dimensions

Human domain	Generative AI role	Positive contribution	Risks and dilemmas	Illustrative example
Identity and expression	Avatar and voice synthesis	Empowered self-representation	Deepfakes, synthetic manipulation	Personalized digital doubles
Creativity and art	Text/image/audio generation	Democratized creativity	Originality and copyright concerns	AI-composed music and literature
Employment and labor	Automated content generation	Enhanced productivity	Job displacement	AI in journalism and copywriting
Communication and dialogue	Chatbots and virtual agents	Accessible information	Miscommunication, emotional harm	AI mental health support bots
Knowledge and education	Tutor bots and writing assistants	Personalized learning	Learning dependency	GPT in academic writing tools
Emotional companionship	AI relationship simulators	Loneliness reduction	Emotional detachment from reality	Virtual friends and love bots
Ethics and morality	Generative policy simulators	Ethical scenario testing	Bias embedding in models	AI debate on moral dilemmas
Media and democracy	AI-generated news and scripts	Faster content creation	Misinformation and polarization	Synthetic news and fake political speeches
Cultural narratives	Storytelling and localization AI	Language preservation	Cultural homogenization	AI translating indigenous stories
Human agency	Decision support systems	Enhanced cognitive reach	Over-reliance on algorithmic judgment	Auto-generated legal and medical advice

surrounding biases that generative models can betray and perpetuate. Consequently, two distinct forms of research inquiry have taken shape around generative AI: first, studies examining how end users can and want to interact with the technology; and second, investigations into designing and training generative models to be less biased, more transparent, and more controllable and explainable [29].

1.7.2 Human-AI Collaboration

While some tasks may be simplified and automated through pre-existing procedures, the creation of new, innovative content typically requires substantial effort and complex decision-making. The same trend is evident in the growing opportunities to collaborate with AIGPT. For instance, a human might initiate a story with a few paragraphs, delegating routine or dull portions to the AI, and subsequently, after reviewing the AI-generated chapters, employ the model to propose potential plot developments. Similarly, in areas like fine arts and programming, humans dictate key decisions that ultimately shape the final creative product, which in both cases derives from human–machine collaboration. Indeed, as observed in user experiences with ChatGPT, humans can leverage generative AI to enhance their work, fulfilling varied functions as assistants, tutors, coworkers, or even friends [30]. This suggests that generative AI will support human collaboration, functioning as either a partner or an assistant. Whether perceiving the generative process as mentalistic, communalistic, or mutualistic, AIGPT-based technology becomes integrated into a system requiring human-induced external inputs to generate a finished product, following an inner real-time feedback loop that collectively connects to all the significant parameters of the final output [31]. Under each feature vector, the strength of human-AI collaboration can be summarized as follows:

(i) Information: Human collaboration is essential to provide input material.
(ii) Creativity: The resulting product reflects a fusion of human and machine creativity.
(iii) Time: Automation of routine and redundant tasks accelerates the process.
(iv) Acceptance: Users adapt to and incorporate generative AI into their creative work.
(v) Complexity: Collaboration in complex decision-making and intermingling tends to improve results.

1.8 Ethical Considerations in Artificial Intelligence

Ethical challenges in AI are now a recurrent theme in debates concerning these technologies. Given the increasing capabilities of generative AI to produce text, images, music, or videos, evaluations of their impacts on work and society more broadly must include ethics. Concerns over potential risks and negative consequences have yielded calls for caution in development and deployment. Indeed, human-centred analyses of AI are not new but have emerged continuously with practical advances and breakthroughs in applications. Examination of ethical considerations encompasses systematic biases in data supplied to models, the discrimination problems demonstrated in AI outputs, such as language translations signalling gender prejudices, and the transparency and accountability—or lack thereof—of today's AI-driven services. Much has been written about AI's potential to deepen societal divisions and inequalities, as well as about potential harms such as disinformation and disempowerment [32]. It is equally important, however, to consider how these technologies may better serve human purposes and facilitate meaningful collaborations with humans.

1.8.1 Bias and Fairness

Bias and unfairness are among the most widely debated ethical concerns in artificial intelligence. AI is not inherently biased; rather, human biases are inadvertently embedded in it through data and code. AI systems often replicate and even amplify the implicit and explicit biases of their creators, especially when trained on unfiltered inputs. Machine learning models are data-driven and statistical by design, making them susceptible to biases arising from imbalanced or incomplete datasets, incorrect variable selection, inappropriate model choice, or flawed training procedures. Even with balanced data, if certain features strongly influence the prediction model, the algorithm may still perpetuate bias. Ensuring fairness in generative AI, therefore, necessitates safeguarding against unfair treatment of diverse groups during data collection and model construction, processing, visualization, and decision-making. Context-specific considerations are crucial because different scenarios entail distinct perceptions of discrimination [33]. However, the opacity and verbosity of most generative models complicate the task of guaranteeing fairness, especially when models learn latent associations within the data.

1.8.2 Transparency and Accountability

Although a precise definition of transparency is elusive, it generally connotes communication and openness. Indeed, AI decision-making that is communicating openly would be viewed as transparent by most people. A transparent AI system is one whose decision-making process can be understood by human users. While transparency fosters user trust through an understanding of AI operations, the term may be inadequate if it does not encompass the full spectrum of trust factors [34, 35]. Transparency is often equated with explainability, yet it is possible for users to feel comfortable with an AI system without necessarily understanding its inner workings; further research into the relationship between explainability and user comfort is warranted. Accountability denotes being responsible for one's actions.

Stakeholders, spanning developers, vendors, customers, and end-users, are tasked with ensuring the responsible use of AI products. Human responsibility prevails, as AI systems remain tools under human control. However, accountability is not confined to those interacting directly with the technology [15]. AI impacts a broad range of individuals, including boxing match opponents affected by AI-driven fight predictions, investors influenced by AI-generated financial advice, and social media users engaged by AI-driven content recommendations [5]. Thus, a comprehensive approach to accountability must consider all parties affected by AI outcomes.

1.9 Impact on Society

Discussions about AI have consistently delved into its effects on humankind, expanding to encompass the creation of autonomous workerless economies, deepening artificial intelligence limitations, the quest for ageless existence, equilibrium among automation and creativity, and implications for healthcare. Societal impact conversations examine AI's influence on individual and communal hierarchies, such as job displacement, social cohesion, long-distance relationships, public funding, social credit scoring, and viral consciousness. Within these debates are focal concerns on bias and fairness, transparency and explainability, and responsibility and accountability. AI-generated creations have raised questions about intellectual property and ownership. These themes further extend into artistic expression and cultural branding. Legal regulations must now address copyright infringements, model development, and ownership of dialog-agent ideas and expressions. Emerging case studies provide insights into generative AI's successes and failures. Media portrayals and educational initiatives shape public engagement with AI [2, 6]. Approaches that integrate expertise from all scientific branches affirm the necessity of an interdisciplinary dialogue—one that weaves together the humanities, artificial intelligence, and information technology.

1.9.1 Economic Implications

Although the benefits of generative AI are yet to materialize fully, very clearly, fears of large-scale job losses are highly unfounded. The generative AI industry, although still in its nascent phase, has generated tens of thousands of jobs, with some of the AI startups becoming unicorns within five years of establishment, with investors projecting the global generative AI market to reach $126 billion by 2030. The notion of AI coding software wi-fi has also been called into question, with studies indicating that programmers spend twice as much time figuring out how to get AI coding software to complete tasks for them as it would take to code from scratch. Virtual assistants such as ChatGPT and Midjourney have become popular among users for their ability to interact naturally with humans and quickly generate various types of content, pointing towards the possibility of hope for mass concentration on more creative positions in the workplace in the future for the first time in history [1].

Moreover, new job positions are being created from the demand for prompt engineers and generative AI trainers. Generative AI enables new experiences such as AI creation, interaction, and distraction, which opens up opportunities for completely new job categories and forms of income. However, in the shorter term, several challenges appear. Generative AI could provide a stronger, more coherent visualization of fake news, including placing deepfakes in multiple contexts. This is particularly dangerous when misleading information about a company's stock affects investing decisions and its compliance with regulations. Likewise, the impact on SEO engines of the propagation of largely shallow content could result in financial hardship for a considerable and very well-established group of people. The way that generative AI tools currently operate also introduces potential biases in the content generation pipeline [28]. Although the magnitude of the effect is starkly unclear, the negative influence of generative AI on information acquisition is an important consideration.

1.9.2 Social Dynamics and Relationships

Social dynamics and relationships are also transforming, influenced by the ways generative AI adapts technology to everyday applications in numerous sectors, including education and the sciences, or even just communication with the outside world. The integration of AI by countless users, regardless of technical expertise, inevitably raises many controversial ethical issues: how can people be protected against the generation of harmful, violent, indecent, or otherwise problematic content? What about frequent cases of bias and discrimination? Is it possible to guarantee that sugary responses accustomed to gaining public approval are not countersigned at any time with the appropriate sarcasm? Even knowing how to live nicely in the shadow of large companies that seek to exploit the technology in the new era risks being an insurmountable task for many individuals [29].

1.10 Legal Frameworks Governing Artificial Intelligence

Governing legal frameworks play a pivotal role in steering generative AI research and shaping its consequences on society. Intellectual property issues and regulatory challenges are two critical aspects of the legal domain that must be examined to ensure the responsible adoption and use of generative AI. Intellectual property (IP) issues pertaining to generative AI involve questions about the ownership of AI-generated content: whether the creator of the AI model, the user who inputs commands, or the AI entity itself holds the copyright. Additional concerns include the use of copyrighted training data and potential infringements in AI-generated outputs. Regulatory challenges encompass the establishment of guidelines for AI use that prevent privacy invasions, biased decisions, misinformation, and fraudulent activities. Such regulations should also guarantee transparency, explainability, accountability, and unbiased outcomes. International initiatives like the European Union's proposed AI Act and the US AI Bill of Rights represent steps toward defining the responsibilities of AI-generated content developers, users, and stakeholders [31].

1.10.1 Intellectual Property Issues

Intellectual property concerns emerge in tandem with the production of creative artefacts by generative AI, particularly during the transition where increased quality enables use in downstream projects. Developed wings for combat aerial vehicles and sanitary fitments might no longer receive extensive analysis were it not for the external attention generated by tools such as ChatGPT. The conversation around intellectual property rights changes accordingly. Questions arise at both ends of the spectrum with respect to copyright protection. On one side stands existing material used, extensively or otherwise, to train generative AI; on the other, the generated output navigating near-perfect convergence. Specifically, emergent questions consider the originality and protectability of AI-generated artefacts—who owns the output? The legal debate soon intersects with related fields and public opinion [36]. For instance, does training a generative model on behind-paywall academic journals amount to theft? Are Wikipedia contributors adequately compensated for their gratuitous labour being exploited to reduce the bill for users' information needs? Could certain models trigger delisting in Google Search due to novelty accusations? The attention sweep broadened further as Gutenberg turned voice, preceding the long-term embedding of grounded knowledge.

1.10.2 Regulatory Challenges

Regulatory challenges associated with generative AI extend beyond those introduced by earlier AI generations. These challenges are linked to the immense power of Web3-generated content, capable of profoundly transforming the realms of culture, economics, society, and politics. Regulation cannot merely address negative aspects; an irrational backlash against generative AI could also suppress its inherent positive contribution to society. Although the European Parliament adopted a legislative resolution on the proposal for LISA in September 2023, the act itself lacks sanctioning or implementation rules and merely outlines the necessary safeguards. LISA should be viewed as a living, evolving proposal. Providing elaborated sanctioning rules and enabling Directive development offer valuable opportunities for societal evolution, permitting discussion, evaluation, and progressive adaptation of AI law and regulation [37]. Consequently, a dark or dystopian Web3 should be neither accepted nor feared, while regulation that encumbers the emergence of a bright and creative Web3 should be strongly contested.

1.11 Trends of Generative Artificial Intelligence

Rapid advances in AI have brought in a new era of AI-enabled applications, products, and platforms in the daily lives of human beings. One category of such AI applications is generative AI. In essence, generative AI systems can create new concepts, ideas, designs, and other forms of creative content based on the dataset that the underlying model is trained on. More generically, they can be considered as an innovation machine. Since its very inception and conception, generative AI has drawn significant attention from human beings. Beyond the fundamental AI or computer science discipline, the very concept of generative AI also closely relates to human life, society, and even the entire world. As such, it is essential to also carefully look into the impact of generative AI on human and societal aspects. The synergy between people and the AI world is equally worthy of attention. The area of Human–Computer Interaction (HCI) is well known for its focus on users and their needs. From the viewpoint of key human considerations, questions arise concerning the generation of large-scale and diverse responses and services to human requests through the utilization of powerful generative AI models. It is critical to explore the challenges associated with the usability of generative AI in the human world. The ethical issues, such as biases, fairness, lack of transparency, and accountability of the models, have already been widely discussed. Inevitably, the study also extends to the economic or social impact that systems like ChatGPT may cause in the near future [38]. A different perspective contemplates the legal aspects that generative AI may trigger, particularly the intellectual property issues conferred by generated content or responses.

1.11.1 Technological Advancements

Generative AI models have been broadly considered advance. Recent technological advances have resulted in large generative models that are, in some tasks, performing at a human level. Generative AI systems have advanced considerably; since 2013, models have been scaling exponentially and performing progressively complex tasks in a wide variety of domains. In creative and artistic generation, generative models have reached a remarkable level of detail: upon receiving a prompt in natural language, they can create highly realistic images, music, video, code, and text. These models have successfully filtered out harmful information and reinforced playfulness and curiosity. They manifest signs of empathy and an understanding of the intentions of other agents. Yet, despite these impressive advancements, fundamental limitations rooted in the models' underlying architecture remain. Models correctly generate information in certain scenarios but fail in unexpected yet simple ways. Researchers continue to encounter various structural flaws resulting in negative consequences, such as reinforcement of biases and inaccuracies, hindering broader societal application [26]. These issues have been addressed mainly by increasing model size, but current generative models require significant enhancements in human interaction to mature. As a result, attaining artificial general intelligence remains a remote prospect.

1.11.2 Potential Societal Changes

Generative AI possesses vast disruptive potential. It might decimate employment across industries and alter people's relationships with money, with each other, and with the nature of work. It has encouraged unprecedented levels of cooperation—especially between competitors—and also a massive arms race. It contributes to feelings of anxiety and depression, while also serving as a source of comfort for many. AI represents humanity's first real chance to consciously design society, as the social norms, expectations, and values surrounding AI will shape the future of humanity itself. After an initial bout of enthusiasm and promise, society is beginning to refrain from making grandiose political promises about these technologies and instead focus on governance, which is far less exciting. When considering and deliberating on AI, societies must assess their appetite for risk and the level of disruption they can or should endure. Failure is likely to occur, but not failure of the technologies themselves, rather failure at shaping these technologies around the inherent strengths and weaknesses of human beings and the societies they create [28]. Although these technologies are extraordinarily impacting, they remain a reflection of humanity rather than a cause for despair.

1.12 Case Studies of Generative AI

Generative AI has become increasingly topical in recent years and is now emerging in all possible social and professional fields. These new digital technologies dedicated to generating content constitute a broad class of technologies based on deep learning aimed at producing new digital content—images, text, speech, video, computer programming—from patterns learned in large volumes of training data. The text generated by ChatGPT in response to a precise query is not derived from a text that it has simply copied; it is an original text that does not pre-exist anywhere else. Understood in this way, generative AI raises many questions in terms of trustworthiness, bias, transparency, and human rights. The specificities of generative AIs, notably ChatGPT, and their consequences in terms of impacts on professional productivity, business models, intellectual property issues, education, and legal responsibilities deserve much attention [32]. These aspects of generative AI are now the subject of more than a hundred different areas of digital work.

1.12.1 Creative Industries

Generative Artificial Intelligence has a wealth of practical applications in creative industries spanning the visual arts, photography, and music sectors. Drawing inspiration from diverse areas such as painting styles and photographic techniques, generative AI is equally capable of producing outputs that are highly imaginative and abstract. Beyond the creative realm, the healthcare field boasts a wide range of tasks where the adoption of generative models has proven beneficial. Applications include drug design, drug discovery and repurposing, molecular optimization and synthesis, molecular property predictions, activity and toxicity analysis, de novo peptide and protein design, scaffold hopping, and de novo RNA design. In other domains, generative models assist in natural language processing applications such as automatic story plot generation and dialogue generation. Applications of Generative Artificial Intelligence extend across the ITC research themes. For instance, in the Awareness theme, the technologies support use cases involving automatic content generation, with necessary services provided by the Governance theme. Within the Human theme, considerations of human-AI interaction examine the relationships between people and machines as new technologies emerge [33]. In the Society theme, analyses focus on how AI is shaping society across various dimensions, including economics, commerce, and regulation. Together, these diverse yet interconnected applications underscore the broad impact of generative AI across multiple sectors.

1.12.2 AI in Healthcare

The advent of artificial intelligence in healthcare was an inevitable next step, given the field's reliance on classification in diagnosis and treatment. Among the earliest clinical decision-support systems was the 1970s MYCIN, which significantly improved the diagnosis and antibiotic-agent selection for certain infectious blood diseases. More recent endeavors include IBM Watson's World Editor project, which parses medical journal articles and could aid researchers in staying abreast of developments. Yet practical clinical applications of AI routinely transcend their symbolic antecedents. Expert systems that prescribe appropriate dosages based on a patient's bowel habits exemplify real-world clinical expertise codified in program form [39, 40]. Equally successful are applications that combine symbolic and nondeterministic processing, process the non-narrative portions of medical charts, or attempt to predict the probability of strokes, analyze arterial blood samples, or identify metastatic lung cancer.

1.12.3 AI in Finance

AI is reshaping the finance industry. According to a report by Autonomous NEXT[4] A significant majority of executives (72%) in investment management acknowledge that AI can significantly impact their business, while an earlier report by Pricewaterhouse-Coopers highlights that 73% of financial institutions are presently implementing or exploring the adoption of AI technologies. These trends are set to drive a large proportion of the expected $22.6 billion market size of AI in FinTech by 2030. Specific applications include algorithmic trading, virtual assistants, fraud detection, credit risk analysis, and high-frequency trading. AI technologies powered by historical and real-time data have evolved to uncover market insights and trends, enabling investors to make more profitable decisions [41]. An example is Civis Analytics, which developed AI-driven systems for stock price prediction. TicToc, an Australian non-bank lender, uses AI to process mortgage applications quickly and more efficiently than traditional lenders. The use of customer and public data allows companies to better assess the level of risk that customers present, overcoming the limitations of historical data, which is often unable to generalize risk levels for academia as well as customers. Experiences with generative artificial intelligence technologies have not always been successful. The public still remembers the retraction of the chatbot Tay in 2016. Among other failures can be numbered chatbots such as Plato, Nightmare, and AI-chan. Failures are also numerous in the domain of creating various types of multimedia using generative AI methods and models [28, 30]. The compromised content generated by generative AI raises ethical and legal questions and may have a negative impact on the public.

1.13 Public Perception of AI

Artificial intelligence has been a popular topic of science fiction for many years, with filmmakers fantasizing about intelligent machines that either assisted or harmed people. Thanks to these works of fiction, the public is forever developing opinions about what AI might be capable of. Every new advancement is either heralded as a breakthrough or feared as the first step towards intelligent machines taking over the world. Sometimes rapid developments lead to increased public awareness and interest in AI, as was the case when programs such as Deep Blue, Watson, Imagen, AlphaGo, GPT-3 3 and DALL-E defeated human players in games, answered trivia questions, or created artificial images and text. During these times, AI-related headlines change from focusing on the social and economic issues of recent times to the opportunities and risks associated with the rapid development of generative AI [5]. The media plays an important role in shaping discussions about AI. Whenever a new advancement is made, it affects residents' awareness and perception of the technology. Careful consideration should be placed on the content of these discussions. Educational

[4] https://www.lexsokolin.com/autonomous-next.

bodies must consider whether the public has the knowledge and skills to evaluate the claims made by entertaining but often overhyped media stories. People should be sufficiently informed to appraise the opportunities and risks of AI and be able to engage in knowledge-driven discussions on the topic. Therefore, those involved in the development of AI must reflect on society's feelings about what it means to be human and the effect that AI will have on human economic activity, interpersonal relationships, social values, and morality [17].

1.13.1 Media Representation

Advancements in AI over the years have garnered significant attention from the general public and communities concerned about its effects on society. This coverage, which often highlights AI failures, highlights potential harms and biases, and explores the possibility of AI developing sentience, plays an essential role in shaping public discourse and directing research efforts to address such concerns. These biases towards danger and dystopia find expression in popular television shows, such as Black Mirror and Westworld, as well as in the characterization of HAL 9000 from 2001: A Space Odyssey. Other media outlets portray the future of AI more optimistically, emphasizing the productive results it can help achieve. Media narratives about AI also inevitably influence policy decisions. Notable examples include New York's numerous bans on facial recognition technology and the European Union's proposal to consider sentient AI systems as legal entities. Despite the considerable potential of generative AI applications, these social and political concerns may impede or distort developments in the field [18]. Scientific and technical media, such as Chemical and Engineering News (CandEN), have not escaped these trends either: coverage of generative AI appeared immediately after ChatGPT's release in November 2022, with a headline emphasizing that "The AI craze is here–but the conversation over risks has barely started." In the interest of providing a broader view, this article will explore a more extensive range of issues.

1.13.2 Public Awareness and Education

The broad adoption of generative AI tools has highlighted the scarcity of teaching and training focused on the general public, adding to other deficiencies in public awareness. An initiative has been applied aimed at different audience profiles, starting with a simple definition of the new concept, as an evolution of traditional AI, the main discoveries underlying its development, the most popular applications, the considerations of the human–algorithm interaction, and finally the ethical, social, and legal aspects accompanying the use of the technology [2]. Work has also been done to address the absence of formal teaching of AI in undergraduate computer science programs, since this is an essential subject for a basic understanding of AI and is lacking in many academic curricula. The result is an openly accessible, easy-to-implement project to be developed by the professor in a single visit to the computer laboratory or remotely. It allows topics such as the building of machines able to generate content automatically, the use of powerful language models, or the training of generative adversarial networks to be introduced. The educational unit is self-contained, utilizes Jupyter Notebook, requires no previous technical knowledge, and is supported by short theoretical explanations based on the research work underlying the tools selected to teach the subject, as some approaches presented in Table 1.2.

1.14 Interdisciplinary Approaches to AI

The development of AI in the broadest sense, any technique that enables computers to mimic human intelligence, has a long history: attempts to build such an intelligent system date back to the mid-twentieth century. Today's interest in generative AI, ushered in by success in natural-language applications, stems from the creation of generative models. These models are capable of utilizing extensive amounts of data and resources to learn the underlying distribution of data, shedding light on the nature of the question. Their applications have traversed creative arts and pharmaceutical development and entered the lives of millions seeking entertainment and insights. Given the profound impact of generative AI technology today, a human and societal perspective is crucial, considering principles, emerging concerns, interpersonal relations, business activities, and economic consequences, together with the legal framework surrounding it. With AI poised to reshape human society, humanities scholars are acutely aware of the cultural ramifications these technologies portend [42]. Such a multifaceted system demands an interdisciplinary approach that brings together ethics, sociology, economics, psychology, law, and foundational advances in technology and AI itself.

1.14.1 Collaboration Across Fields

Collaboration among technical and social disciplines affects every stage of generative AI research, from elaborating the foundational concept, to its possible applications, and to its consequences for human–AI interaction. Interdisciplinary studies can help mitigate any new human challenges for the medium, improving user experience via better design. They also enable the exploration of the complex social dilemmas that arise as AI production enters the cultural sphere. Without integration, ethical values can be hard-coded uncritically into a non-representative agent or robot, and without mechanisms for accountability, privacy, and security, negative individual or societal repercussions remain very real [18]. Using an economic, social, and political angle, with consideration given to specific legal rules, is also advised when analyzing future trends in generative AI.

Table 1.2 Societal systems and frameworks engaging artificial intelligence

Societal system	Application of GenAI	Systemic transformation	Governance challenges	Emerging response models
Education system	AI-Generated curricula	Personalized, adaptive learning	Plagiarism, intellectual ownership	Ethical AI literacy programs
Health and wellness	Symptom-to-text generators	Efficient diagnostics	Medical misinformation	Explainable AI in clinical decision tools
Governance and law	Policy drafting models	Accelerated legal document review	Legal ambiguity of AI outputs	AI accountability frameworks
News and journalism	Article writing agents	24/7 Reporting infrastructure	Truth versus machine bias	AI Fact-checkers + human oversight
Art and culture	AI-curated exhibitions	Participatory creativity	Cultural erosion	AI-human co-creation protocols
Finance and economy	Automated report writers	Faster decision cycles	Manipulative market influence	AI disclosure in financial reporting
Religion and philosophy	AI sermon and text generation	Multifaith access and insights	Theological misrepresentation	Human-in-the-loop ethical boards
Surveillance and security	Deepfake detection systems	Authenticity validation	Privacy and civil rights erosion	Algorithmic auditing agencies
Environment and climate	Eco-narrative generators	Mobilized green storytelling	Tokenization of nature	Generative climate advocacy platforms
Civic participation	Policy simulation interfaces	Data-informed public engagement	Elite-coded narratives	Civic-AI collaboration portals
Humanitarian systems	Crisis scenario simulators	Emergency preparedness	Dependency on synthetic data	AI for humanitarian aid coordination
Labor and gig economy	Auto-gig matching algorithms	Just-in-time employment	Disempowerment of workers	Transparent algorithmic labor rights
Scientific discovery	Hypothesis generation tools	Accelerated research innovation	Replicability and fabrication	AI validation and peer-review assistants
Family and relationships	Digital companionship platforms	Support for emotional isolation	Authentic connection versus illusion	Co-regulation with human-centered design

By reviewing milestone events and important breakthroughs in the historical development of AI, the area of generative models can be presented within a clear narrative. Although it remains a human-created human tool, so that society determines the purpose for which the technology is eventually put to use, misunderstandings or misapplications of generative AI might still alter the perception of the related field. Case studies of successes and failures, therefore, can be incorporated into the discussion. Exposure to popular media usage, including various entertaining classification systems, is yet another valid lens when assessing public representation, knowledge, and awareness [42]. Thus, advanced training must be readily available in a two-way information flow: humanities scholars should possess sufficient technical knowledge to keep up with the rapid development of AI tools, while STEM researchers will greatly benefit from adding ethics, sociology, or history modules to their curricula.

1.14.2 Integrating Humanities and Technology

It is particularly exciting to contemplate AI's next phase, in which the field treads beyond the familiar. Questions inevitably emerge regarding the alignment between the intentions of builders and the outcomes of their creations. Accordingly, practitioners from both the technical and humanistic domains are motivated to explore these concerns. Businesses, too, feel the pressure to formulate new policies that protect users. Regulations are deliberated at the nation-state and company-action levels. Scholars in philosophy, religion, law, and economy, among other disciplines, contribute to the discourse as well. The present special collection aspires to showcase these very perspectives, views that humanize AI, instead of humanizing AI as portrayed in fiction. During the final phases of editing, generative AI systems continued to make headlines, notably ChatGPT-4, which supports both visual and audio inputs. Further developments in all capacities are anticipated in the near future, reconfirming that generative AI is the new business trend. However, beyond the apparent industrial interests, efforts towards humanizing AI and considering its impact on humanity remain invaluable [4]. Shaping perspectives in this manner safeguards society from the consequences of unregulated innovations.

1.15 Future of Artificial Intelligence

AI has rapidly evolved from an academic pursuit into an integral part of modern life, with a trajectory still unfolding. Looking ahead, predictions for AI's future encompass areas such as deep learning, everyday applications, and ethical considerations. AI systems are expected to attain levels capable of performing a wide range of daily activities more effectively than humans. This progression is fueled by data-science applications, which learn and improve from extensive datasets, kludging day-to-day tasks in domains as diverse as healthcare, finance, and life insurance. However, even as AI approaches superhuman levels on numerous tasks, substantial challenges remain in tackling more complex problems [1]. Looking toward potential disruptions, AI could lead to employment displacement and ecological impact alongside job creation and resource efficiency. Although broad societal concerns continue to focus on the control of massive, superintelligent AI, the immediate disruption lies with concrete, short-term slicers: concentrated models disrupting white-collar jobs or companies dominating downstream economics. A key emerging area of concern is that of regulating the capabilities of AI. The increased capabilities embodied in large language models and emerging techniques to steer them have raised fears of dangerous, rapid advancements in AI. Both companies and governments are seeking ways to regulate or slow such developments, drawing parallels to the regulation of more dangerous areas of research, such as genetic engineering [5].

1.15.1 Predictions and Trends

In 1959, economist Herbert Simon made an intriguing prediction that foresaw a world deep in the throes of data automation: "machines will be capable, within twenty years, of doing any work a man can do." It was a hopeful forecast, and, being an economist and a sociologist, Simon embraced the Enlightenment vision of progress. Although his 1973 Nobel Prize did not pertain to work in artificial intelligence, used cautiously the statement captures the mood of the read at the time. Some fifty years on, things are moving in the direction he described. || In 2017, the concept of the "Future of Anything" became a hot topic among artificial intelligence writers and marketers, not to mention executives overseeing a spectrum of business interests. Whether the subject is "self-driving cars," "virtual medical assistants," or "automated financial advisers," it is significant that both the media and its audience recognize the affinity these topics share with the intelligent machines of culture and story. Still, it seems that the considered judgment of the economics professor of 1959 was ambitious [29]. When the announcement of Google DeepMind's AlphaGo victory was met with terms such as "astounding," "sneaky," and "far-future AI," it was clear that progress had yet to fulfill Simon's prophecy.

DeepMind has neither developed a driverless car nor solved automated medical diagnostics; it made a specialized product that performs a specific routine better than a single human exposure to the problem. This observation forms part of a more recent narrative offered by a writer for The Atlantic: "When it comes to the future of A.I., the company has a more guarded outlook than the one offered by the media." Behind the excitement of the newsstands, a bigger question, profitably posed within the business-media context: "Can A.I. Deliver on Its Ever-Growing Promise?" Its myriad applications hardly obscure the theoretical fact that current implementations represent the product of a haunted dream, one whose source can be traced back to Claude Shannon's observation of a chess-playing machine and Alan Turing's realization that the sequential control of a universal machine would permit it to play an infinite number of games [32].

1.15.2 AI and Human Collaboration

Human senses constantly collect data and integrate it into consciousness... An AI can surpass human capabilities, handling many tasks alone. However, combining human intelligence and AI power creates optimal synergy one led by AI, the other by human capacity. Human collaboration emerges as a vital resource for methods requiring insights beyond AI's autonomous purview. Using intelligent software with a patent database illustrates AI and human collaboration. AI-generated patent claims that exclude prior art nonetheless require human review to identify false or misleading claims. Humans play a crucial role in building knowledge and logic within intelligent software, exposing it to novel experiences. In each case, AI and human strengths interweave seamlessly.

1.15.3 Regulatory Challenges

Debates around AI regulation have escalated sharply, with many calling for protective measures against the risks posed by increasingly powerful systems. The immediacy of the threat is debated—some argue that despite current rapid progress, a robust regulatory framework requires more lead time than when the dangers become glaringly evident. Others classify these risks as existential and pressing enough to warrant immediate action, with calls in the US for a five-year pause in AI development once the technology attains human-level competence. Different regions have adopted distinct regulatory approaches. The European Union's comprehensive AI Act envisages harmonizing rules to support cross-border AI adoption while enforcing human-centered standards. In contrast, the United States has relied more on principles, emphasizing equitable, secure, and transparent AI through advisory bodies. A significant challenge lies in the rapidly evolving nature of AI technologies, which advances faster than laws can adapt, potentially rendering regulations obsolete shortly after implementation [33]. Moreover, effective AI governance demands unprecedented international cooperation to manage risks beyond national borders.

1.16 Conclusion

Artificial intelligence, the simulation of human intelligence in machines that are programmed to think and learn like people, has captured the collective imagination for decades. It began as the still, inanimate machine suddenly stirring and waking, shaking off its first shroud of logic, and beginning to envision for itself a prophetic future. Today, it has awakened fully: actively imagining and realizing its future. AI is no longer a technical curiosity, but an inseparable feature of modern life. Already, AI applications have moved beyond research and theory and are making a practical, visible difference in the world. The deep learning techniques that drive AI provide an almost unlimited variety of creative applications, capable of improving performance and lives across broad swaths of daily practice. It will be most exciting to see where the machine decides to go from here, as it creates an increasingly reciprocal partnership with humankind. The key news in Artificial Intelligence is neither technological nor technical but is about humans and, through humans, about society. This shift in perspective is well established in other disciplines related to AI (for instance, ethics) and in the social sciences and humanities, but not yet in the fields of machine learning, deep learning, and neural networks. An introduction to Generative Artificial Intelligence oriented to humans and society provides a contemporary and comprehensive historical background. The basic notions that underlie Generative Artificial Intelligence are explained, followed by a description of representative use cases. Then, the human interaction with Generative Artificial Intelligence is briefly sketched, and finally, the ethical, societal, and legal aspects.

References

1. Scott Hansen S. Public AI imaginaries: how the debate on artificial intelligence was covered in Danish newspapers and magazines 1956–2021. Nordicom Review. 2022;43.
2. Kaluarachchi T, Reis A, Nanayakkara S. A review of recent deep learning approaches in human-centered machine learning. Sensors. 2021;21.
3. Rezaev AV., Tregubova ND. ChatGPT and AI in the Universities: An Introduction to the Near Future. Vysshee Obrazovanie v Rossii. 2023;32(6).
4. Wang L, Zhang Z, Wang D, Cao W, Zhou X, Zhang P, et al. Human-centered design and evaluation of AI-empowered clinical decision support systems: a systematic review. Front Comput Sci. 2023;5.
5. Xu W, Dainoff MJ, Ge L, Gao Z. From human-computer interaction to human-AI interaction: new challenges and opportunities for enabling human-centered AI. Int J Hum Comput Interact. 2021;39(3).
6. Rony MKK, Kayesh I, Bala S Das, Akter F, Parvin MR. Artificial intelligence in future nursing care: exploring perspectives of nursing professionals—a descriptive qualitative study. Heliyon. 2024;10(4).
7. Ellingrud K, Saurabh KE, Gurneet S, Dandona S, Madgavkar A, Chui M, et al. Generative AI and the future of work in America. McKinsey Global Insitute. 2023.
8. Stahl A. How AI will impact the future of work and life. Forbes. 2021.
9. Jarrahi MH, Lutz C, Boyd K, Oesterlund C, Willis M. Artificial intelligence in the work context. J Assoc Inf Sci Technol. 2023;74(3).
10. West DM. The future of work: Robots, AI, and automation. 2018.
11. Banafa A. AI and the future of work. In: Transformative AI. 2024.
12. Oniani D, Hilsman J, Peng Y, Poropatich RK, Pamplin JC, Legault GL, et al. Adopting and expanding ethical principles for generative artificial intelligence from military to healthcare. NPJ Digit Med. 2023;6(1).
13. Manyika J, Sneader K. AI, automation, and the future of work: ten things to solve for. McKinsey Global Institute. 2018.
14. Shafik W. Deep Learning impacts in the field of artificial intelligence. In: Deep Learning Concepts in Operations Research [Internet]. New York: Auerbach Publications;2024. p. 9–26. https://doi.org/10.1201/9781003433309-2.
15. Paraman P, Anamalah S. Ethical artificial intelligence framework for a good AI society: principles, opportunities and perils. AI Soc. 2023;38(2).

16. Satornino CB, Grewal D, Guha A, Schweiger EB, Goodstein RC. The perks and perils of artificial intelligence use in lateral exchange markets. J Bus Res. 2023;158.
17. Carriço G. The EU and artificial intelligence: A human-centred perspective. European View. 2018;17(1).
18. Bond RR, Mulvenna M, Wang H. Human centered artificial intelligence: weaving UX into algorithmic decision making. In: RoCHI 2019: International Conference on Human-Computer Interaction. 2019.
19. Shafik W. Introduction to ChatGPT. In: Advanced Applications of Generative AI and Natural Language Processing Models. 2023.
20. Shafik W. Generative artificial intelligence for social good and sustainable development. In: Generative AI: disruptive technologies for innovative applications. 2025;153–85.
21. Shafik W. Generative AI for social good and sustainable development. In: Generative AI: current trends and applications. Springer;2024. p. 185–217.
22. Bühler MM, Jelinek T, Nübel K. Training and preparing tomorrow's workforce for the fourth industrial revolution. Educ Sci. 2022;12.
23. Shafik W. Toward a more ethical future of artificial intelligence and data science. in: the ethical frontier of ai and data analysis [Internet]. IGI Global;2024. p. 362–88. https://doi.org/10.4018/979-8-3693-2964-1.ch022.
24. Shafik W. Generative adversarial networks: security, privacy, and ethical considerations. In: Generative artificial intelligence (AI) approaches for industrial applications. Springer;2025. p. 93–117.
25. Shafik W. Ethical and legal considerations in artificial intelligence. In: AI-Enabled threat intelligence and cyber risk assessment. 2025;90.
26. Kompella K. The trolley problem and ethical dilemmas in AI. Inf Today. 2020;37(5).
27. DataEthics. Addressing ethical dilemmas in AI: listening to engineers. Association of Nordic Engineers. 2021
28. Gouvea JS. Ethical dilemmas in current uses of AI in science education. CBE Life Sci Educ. 2024;23(1).
29. Vousinas GL, Simitsi I, Livieri G, Gkouva GC, Efthymiou IP. Mapping the road of the ethical dilemmas behind artificial intelligence. J Polit Ethics New Technol AI. 2022;1(1).
30. Guan H, Dong L, Zhao A. Ethical risk factors and mechanisms in artificial intelligence decision making. Behav Sci. 2022;12(9).
31. Jabotinsky HY, Sarel R. Co-authoring with an AI? Ethical dilemmas and artificial intelligence. SSRN Electron J. 2022.
32. Prikshat V, Patel P, Varma A, Ishizaka A. A multi-stakeholder ethical framework for AI-augmented HRM. Int J Manpow. 2022;43(1).
33. Zhang Z, Chen Z, Xu L. Artificial intelligence and moral dilemmas: perception of ethical decision-making in AI. J Exp Soc Psychol. 2022;101.
34. Shafik W, Singh R, Kumar V. Artificial intelligence transparency and explainability in sustainable healthcare. In: Transforming healthcare sector through artificial intelligence and environmental sustainability. Springer;2025. p. 165–91.
35. Singh R, Shafik W, Crowther D, Kumar V, editors. Transforming healthcare sector through artificial intelligence and environmental sustainability [Internet], vol. 1, 1st ed. Singapore: Springer Nature Singapore;2024. https://doi.org/10.1007/978-981-97-9555-0.
36. Machado H, Silva S, Neiva L. Publics' views on ethical challenges of artificial intelligence: a scoping review. AI Ethics. 2023.
37. Casas-Roma J, Conesa J, Caballé S. Education, ethical dilemmas and AI: from ethical design to artificial morality. In: Lecture notes in computer science (including subseries Lecture Notes in artificial intelligence and lecture notes in bioinformatics). 2021.
38. Nassar A, Kamal M. Ethical dilemmas in AI-Powered decision-making: a deep dive into big data-driven ethical considerations. Int J Responsible Artif Intell. 2021;11(8).
39. Shafik W. SDG 3: Good health and well-being—digital health solutions. 2025. p. 135–61.

40. Shafik W. Connected healthcare—the impact of internet of things on medical services. In: Artificial intelligence and internet of things based augmented trends for data driven systems [Internet]. Boca Raton: CRC Press; 2024. p. 181–217. https://doi.org/10.1201/9781003497318-10.
41. Johnson J. The AI commander problem: ethical, political, and psychological dilemmas of human-machine interactions in AI-enabled warfare. J Mil Ethics. 2022;21(3–4).
42. Capel T, Brereton M. What is human-centered about human-centered AI? A map of the research landscape. In: Conference on human factors in computing systems—proceedings. 2023.

Chapter 2
The Double-Edged Sword of Artificial Intelligence

2.1 Introduction

The idea that humans have long tried, unsuccessfully, to make entities similar to themselves has been explored in various forms. The ancient Greeks, for example, tried to build human-like entities. The Talmudic legend of the golem also enters this field, while the Pygmalion myth was connected to the speculation that inanimate entities like statues could be animated at will. We can find ideas about artificial entities being created to help humans in both the ancient past and various futuristic utopias and dystopias [1]. It is also evident in more recent American traditions, from tales like the film Metropolis to 1960s television serials such as The Jetsons or Lost in Space. If we look further back, it is equally apparent in Mary Shelley's Frankenstein, using a radically different theme and representation. We should remember that this is a novel, not a construction, as some proponents of second-order cybernetics and AI write. Unfortunately, for the serious development of AI in many respects, this tendency to make entities similar to oneself carries on. Some even go further and ask if these entities could be more functional and more intelligent in order to be useful for people or take the place of people in different tasks [2].

A step towards AI took place in 1997 in a famous game when a machine defeated the then-world chess champion. The game of chess is a recursive finite-game perfect-information tree game. Artificial intelligence agents and logical agents are programmed in accordance with a set of logical assertions specified in a logical language. Such a language has certain syntax and semantics. We refer to the logical agent specifications, expressed in the logical language, as the logical affirmation specification [3]. When an AI agent is given a command, it executes one of five core AI processes, namely, the perception, evaluation, action, goal appraisal, or goal selection process. The processes are executed in tandem and called by either the logic-based intelligent agent authority or the rule-based intelligent agent authority. The perception process produces various percepts about the environment, such as beliefs, desires, intentions, beliefs of others, goals of others, and others' intentions.

W. Shafik, *The Dark Side of AI*, https://doi.org/10.1007/978-3-032-09130-7_2

The belief list is updated as new percepts are obtained. The evaluation process checks the consequences of the action [4].

In contrast to more singular AI, ANI is used to refer to systems designed for specific tasks. These are systems that specialize in valuable troubleshooting capabilities, such as listening, learning, timing, coordinating, commanding, observing, memorizing, visualizing, generating, brainstorming, producing, and reading. They are capable of processing a colossal amount of data and producing information in a form that needs to be interpreted to understand context, value, and correlation [5]. AI is used on data, internet connectivity, and immersiveness. When an agent entity is aware, it is capable of thinking critically. Critical thinking can be broken down into five constituent thinking types, including logical, analytical, metered, creative, and reflective [5]. Cognitive agents can develop and demonstrate executable reflection capabilities, using the current expert systems, presented in Fig. 2.1.

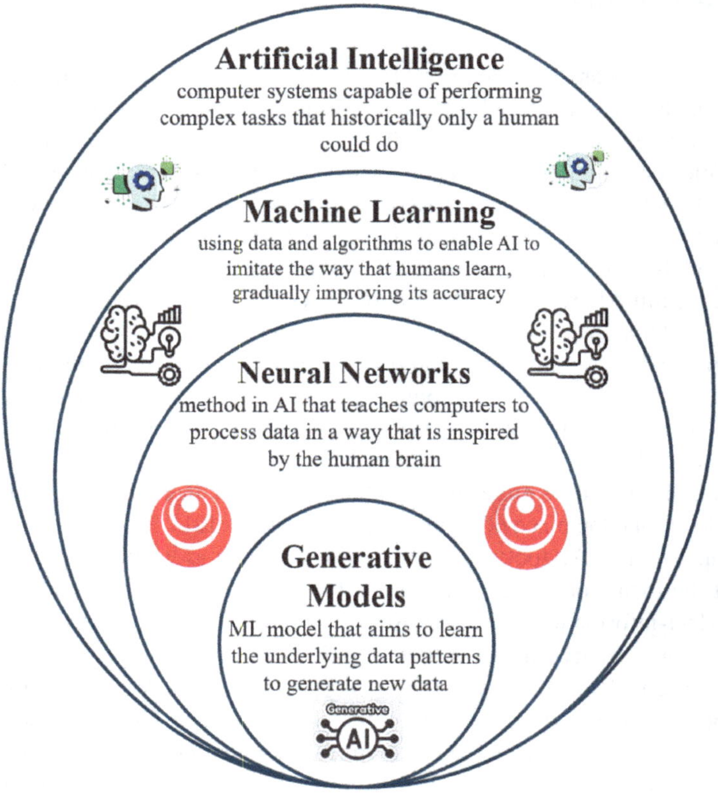

Fig. 2.1 Expert system

2.2 Key Technologies in Artificial Intelligence

AI was developed in the 1950s in order to search for ways of developing computers with the ability to solve difficult problems. After the first summer conference on AI in 1956, AI launched a further search for a more intelligent machine to solve problems from its creators, computer programs, and mathematicians in the 50 s and 60 s, while utilizing other technological components [6]. The result of this search was the invention of the first AI key technologies, such as programs in automatic language processing to control complex objects and interactive virtual agents, developed within a project. After this beginning, the 80 s and 90 s gave us a huge escalation of AI systems and technologies, including a knowledge-based system, parallel hardware to speed up computer systems, more capable LISP software from complex AI, and more efficient computer systems, architectures, and frameworks; genetic algorithms, neural networks, symbolic equations, and formal systems, at least four simulation-based technologies, many fuzzy logic innovations, and AI decision support systems for procurement, corporate acquisitions and mergers, and financial planning. AI has developed a great deal and moved on since the days of its early research and development, despite advancements in big industry, AI government, and its growth is just beginning [7]. During the next few decades, AI will be improved by its infrastructure, which is very basic for its storage and processing of data, at its capacity to accommodate sensorial elements and respond to or search for events in detail, increasing its quantitative and qualitative complexity.

2.2.1 Machine Learning

In the past several years, machine learning algorithms have become perhaps the fastest-growing area of artificial intelligence. Machine learning applications have been developed to recognize just about every type of object, including digital images, sounds, and human expressions. Importantly, machine learning models nowadays can easily be learned to mimic human reasoning [8]. Machine learning models can be trained to help scientists analyze the enormous quantity of scientific data that is generated every day. Models of disease prediction and early detection, cancer biology research, neurological and psychological problems, environmental monitoring, climate change, and conservation modeling are among the successful applications designed for using machine learning techniques [9]. In practice, we encounter machine learning models every day: personalized recommendation services designed in shopping apps, prediction tools for stock and housing prices, image recognition models that help social media automatically categorize digital photos, predictive text generation in email or note-taking applications, online streaming, price-setting models in shopping and transportation applications, autonomous driving cars, and personalized smart home assistant systems designed using sophisticated machine learning algorithms. It is not difficult to build and train models like these, thanks to the vast number of tutorials, learning courses, and programming libraries available. Usually, to design a machine learning model, a computer program needs to be trained for a specific objective by processing large amounts of data. Data are split into two separate components. The most common approach uses a large amount, usually around 80% of input data, to train the model, where the training model selects an example input–output day-to-day pair and sorts the design [10]. The program is initially developed using the entire training data set with the observed inputs and outputs. After the initial design phase, the program is called a "trained model" and can output predictions for new examples that it has not been trained on or has not even been observed at all [1]. The primary aspect of a trained model is that, for prediction tasks, the accuracy of a trained model depends on the quantity, accuracy, and representativeness of its training data.

2.2.2 Natural Language Processing

Natural Language Processing (NLP) is a field of computer science, AI, and linguistics that focuses on the interactions between computers and human languages, particularly how to program computers to process and analyze large amounts of natural language data. With the growing amount of data generated in the digital world every day, NLP systems have been used to perform a variety of language-based analyses. These include sentiment analysis of social media activity or customer interactions for monitoring reputation in the digital environment, named entity recognition for processing vast volumes of written documents, and natural language text processing for transforming text into structured data, or speech-to-text translation and vice versa [11]. NLP models are a double-edged sword since they rely on pre-trained data and models that can be outdated or contain prejudice. If the wrong input is entered into the model, the wrong output is created, which is harmful. Due to the nature of these models being trained on large amounts of data, if they are not given adequate information, the model will obviously fail. This could lead to real-life consequences, as these models are capable of doing a wide range of tasks. Leaders in fields such as healthcare should strive to understand and communicate about the limitations of language models so society can adopt precautions to avoid unwanted, damaging outcomes [12].

2.2.3 Computer Vision

The application of artificial intelligence-based solutions to the interpretation of visual media is often referred to as computer vision. Today, there is a wide range of advanced technologies that are derived from the field. In the space of legal risk, these technologies can explicate video and audio footage, process and transmit customer and business gestures as data, and are commonly used to maintain physical and digital security. Unlike human cognition, AI systems can be trained to perform a specific task by observing a large number of examples. As a result, computer vision is removed from the context of commercial applications and often made responsible for aspects of harm prevention and security. Video surveillance has rapidly infiltrated our urban landscapes, gathering data on millions globally [1]. Video footage is analyzed at data centers to spot unusual behavior, flagging such content in real time. Other common legal applications of computer vision include: the ability to operate transportation that is publicly facilitated; use in automating security; and the control of societal behavior, including in geopolitically significant areas. Platforms with computer vision AI capabilities have automatic deployment systems built in, which can locate weapons in user-generated content from remote places where local authorities are not able to reach [2]. These platforms can also monitor different people's speech and animations, then determine which ads they will be most likely to use to profit from a specific company's advertising.

2.3 Benefits of Artificial Intelligence

Artificial intelligence technologies have had many accomplishments in recent years. For example, a system is helping oncologists determine better ways to treat their patients. A company has developed a self-driving car that they believe is closer to reality each year. An AI chess-playing computer beat world champion Garry Kasparov in a series of matches. A computer program that plays the board game Go surpassed human abilities. AI-empowered knowledge workers are augmenting their decisions. These few examples are just the tip of the AI iceberg. New applications are continually being unveiled that are helping organizations reduce costs, improve efficiencies, and increase their competitive advantages. Clearly, AI technologies have the potential to be a great boon to human society [3]. It is because of these many benefits that AI technologies can bring to organizations that business students need to begin to develop an understanding of what AI is, what it can do, and what it cannot do. AI technologies have the potential to greatly enhance the value of the business intelligence solutions currently utilized in organizations. Companies that are able to gain insights a little quicker, a little deeper, or a little more accurately than their competitors can realize significant competitive advantages. AI technologies have the potential to make these insights a lot quicker, a lot deeper, and a lot more accurate. As a result, it is much more likely that AI will become the next standard addition to an organization's capabilities [13]. With these powerful potential benefits available, business students must become informed so that they can assist their future employers in making wise advancement decisions.

2.3.1 *Increased Efficiency*

Technology is characterized as efficient if it reduces the amount of the factor of production used to perform a given task without causing a reduction in the amount of the task performed. More specifically, a technology is said to be efficient if it permits a given amount of a good to be produced with less of the factor of production per unit of good. New technologies expand the production possibilities of an economy, shifting the economy's PPF outward. What differentiates technological change from other factor input shifts in economic modeling is that the PPF shift occurs without altering the output mix. In other words, the share of national output for different sectors does not necessarily change as a result of new technology. The same amount of resources produces more goods and services, augmenting that nation's standard of living [14]. In the context of technological change, the term "efficiency wage" may well be interpreted in terms of the individual tasks that a worker has been hired to do. A worker who arrives each day, works diligently, and is also cooperative with colleagues contributes significantly to the efficient operation of the firm. If new technology enhances productivity relative to a given performance level, some workers could become redundant immediately. A simpler example would be in the case of discrete manufacturing, when increased productivity from cutting machine technology reduces the number of required operators. While the firm no longer needs as many workers to maintain the same level of production, there are gains to production efficiency in the form of enhanced time of completion of the task that can be expected for those workers who remain [15].

2.3.2 *Enhanced Decision Making*

Artificial intelligence can enhance decision-making. AI systems can process and analyze massive amounts of data much more quickly than humans can. They can simulate and test potential future outcomes efficiently. AI can support optimal decision-making by highlighting probable issues and trade-offs, and by amassing and presenting all available options. This not only makes decision-making more efficient and cost-effective, but also permits wiser decisions. Additionally, well-designed AI assistants can provide decision-makers with increased expertise via an understanding of best practices and lessons learned in numerous similar situations in multiple businesses and industries [16]. They can also provide rigorous rationale and systematic documentation by utilizing a grid-based approach to making complex decisions. This can be important for businesses and regulatory compliance. It is important to note that AI systems could incorporate the impact of decisions on many factors of concern and pinpoint the best trade-offs when values conflict. Such decisions could be more ethical than those made by humans, who may let biases and prejudices cloud sound ethical judgment [17]. Overall, decision-making enhanced by AI can produce better quality decisions, especially at a time when the confluence of accelerating technology

Fig. 2.2 Sampled application of expert systems for human good

change, a growing amount of data, and increasingly connected and interdependent global systems increases the interrelationships among people, organizations, and society, among other advantages as presented in Fig. 2.2.

2.3.3 Improved Customer Experiences

The potential for using details about customers in order to personalize not just to offer discounts to an airline or hotel, but to personalize the entire experience, bodes well for increasing the consumption of various goods. Another type of improved customer experience is to combine products and services in new ways, more effectively. For example, suppose a retailer has information about the site of a factory. In that case, it can often bundle a service contract that provides near-immediate delivery of any replacement parts and a technician trained to install them. This is typically also a win for the customer because the portfolio of related services is then more convenient to use. Customers who are more informed about the services that they have bought and use them more effectively will often have a significantly improved experience [2]. These services don't have to involve the use or consumption of products directly; for example, a bank might use customer data to create budget recommendations based on their goals. It can be thought of as a product that is 100% personalized to the customer.

Having a very rapid, personalized, and deep understanding of a customer, however, can also be unsettling. Companies can get so good at using not only individual transactional data but also the sentiment of communications between the company and the individuals and the feedback they receive at various times that they appear to unduly pressure consumers into keeping relationships with small firms that do not have systems from becoming serious competitors. In their quest to maximize sales, companies often promise more than they can deliver and maintain a high level of ethics with consumers. All these considerations taken together (as well as others) might perhaps necessitate regulation to ensure uniform and ethical service levels across competitors [18]. On the other hand, very personalized offers and recommendations can also be used to increase the customer's satisfaction, which might be yet another reason why they will want to leave early services.

2.4 Risks and Challenges of AI

The great technological, social, and economic promise of the ongoing AI boom is shadowed by potential downsides. This dualism shapes the nature, context, and direction of AI-related technological, policy, and beneficial developments. The better these downsides are understood, the more human society is likely to benefit from the technology, solve potential problems, and avoid major risks and mistakes. The effort will require co-creation by AI and society. Key benefits and concerns will be developed with the help of a social formula of Council-AI-Scientist-Industrial-User Governance. In engaging with development-oriented AI, it is imperative to consider the formidable risks and challenges involved [19]. It is possible to simplify this topic by dividing the matters it encompasses into objective and subjective risks and challenges. The basic taxonomy is as follows. On the AI side, it is necessary to address the risk of AI undertaking being delayed or even aborted, of its potential being inhibited, blocked, distorted, or misappropriated, and of its applications being undesirable or unfeasible. Society, in turn, will be faced with the risk of being unable to capitalize on or reap the benefits of AI, being subjected to undesirable scenarios of AI development or negative impacts of its use, being left unprepared for possible societal problems associated with AI-induced technological change, remaining confined to a too narrow scope of AI capabilities, and, overall, being incapable of dealing adequately with AI and knowledge-based economy-enabled challenges of the future [20].

2.4.1 Job Displacement

The downside of improving technology is that the value of some human labor becomes worth less, and people's earnings decline. In particular, people who do jobs that AI systems can do better or cheaper than humans will feel the pain of AI's arrival. There are several factors people should consider when thinking about job displacement. AI systems are very different from previous waves of technological change, such as those brought about by the advent of machinery or the creation of automobiles. Robot automation is replacing people everywhere, and not just in factory settings. Unlike robotic automation, AI systems can readily perform a very large number of tasks that humans do not perform in a set manner using a tool. Reduced costs coupled with increased supply should lead to more usage of the technology [9]. Over the long term, the technology will discover and stimulate new demand from customers. AI technology will thus provide new usage, more employment, and higher wages for workers who do jobs that conventional AI systems still can't handle.

But it is not all economic sunshine. An economy generally takes some time to move from reliance on an old technology that is labor-intensive to reliance on a new technology that is labor-saving. With AI, the movement might take decades, based on speculation about offshoring's effect on lost U.S. engineering jobs. AI technologies will ultimately profit many creative, skilled workers who develop and train AI systems. The exponential progress of AI is sure to create new opportunities before the economy completes the shift. But throughout the structural transition, many people's jobs will be disrupted. If the recent past is a guide, most of these people will not easily find the new opportunities and will likely be located on the lower end spectrum of income, which will lead to a period of unemployment or personal distress [1]. And for those who do find new jobs, the undoubtedly new skills demanded by future AI technologies might be difficult and costly to acquire.

2.4.2 Privacy Concerns

One of the main immediate concerns that some futurists have raised about new technologies comes from the potential to infringe privacy in unprecedented ways. To some extent, the concerns about the unprecedented nature of modern privacy concerns are correct. At a minimum, from the earliest recorded history until the late nineteenth century, the overwhelming majority of human beings lived in small rural villages where everybody knew everybody else's business. Now, with new data mining and genome decoding companies, an individual's secrets are a potential gold mine. The search algorithm is a fine way to expand people's knowledge about numerous topics; the only problem is that these capabilities come at the expense of potentially violating individuals' desires for the strictest personal secrecy about some subjects, which has long been essential for successful mammal life [3]. Some religious advocates observing the power of the new data mining technologies argue that people have a fundamental "right to be forgotten." We would argue rather that a right to be selectively remembered is essential to mammalian evolution. One reason is that the mammalian life of all of these social primates is, to some extent, boiled down to secrecy. At an individual level, virtually all humans enjoy spending varying amounts of relatively risk-free privacy time in bathrooms and bedrooms, during confession and individual therapy sessions, in business meetings, and during elections [19]. At a societal level, religious sin-eating and/or confession and professional mental health programs provide relatively safe sponges for the most aberrant social "sins."

2.4.3 Bias and Discrimination

AI systems are only as good or fair as the data they are trained on. If the data is biased or fundamentally unfair, then the AI system's output will be too. Furthermore, the use of AI algorithms in decision-making can result in inherent systematic bias based on protected characteristics. AI may thereby be discriminatory against certain groups of individuals. In the context of employment, the use of AI in recruitment tools is being scrutinized on the grounds that it could lead to discrimination. The nine grounds of discrimination covered by the Equality Act in the UK are an area of particular concern. Title II of the Civil Rights Act of 1964 gives people the right not to be discriminated against on the grounds of their color, race, sex, religion, or national origin [21]. National origin could be derived from someone's registered ethnicity. Similarly, the Gender Recognition Act gives a person the right to be recognized in their chosen gender, while the protection against discrimination on the grounds of sex also includes pregnancy and maternity. An entirely unintentional consequence of using AI, especially since the authors are increasingly staring at their models, is the chance that names appearing in data could influence the outcome of an algorithm. Using an applicant tracking system in HR systems could potentially lead to discrimination against a gender-identified person by a male or someone of another ethnic group [22]. This is another example of where the technology is new but the ethics are not, and the biggest section of disabilities is not fully catered for as we transition to a global, sustainable future, as illustrated in Fig. 2.3.

2.5 Ethical Considerations in AI

Ethical discussions about ethical considerations in AI are distinct from discussions about general AI and have an impact on public perceptions of that eventuality. Ethical discussions that are associated with early AI generally fall into two distinct categories. The first consists of specific ethical challenges that AI may raise or exacerbate, or help us mitigate. Such topics include, for example, the possibility of unconstrained development of AI in the absence of agreed ethical safety principles. The

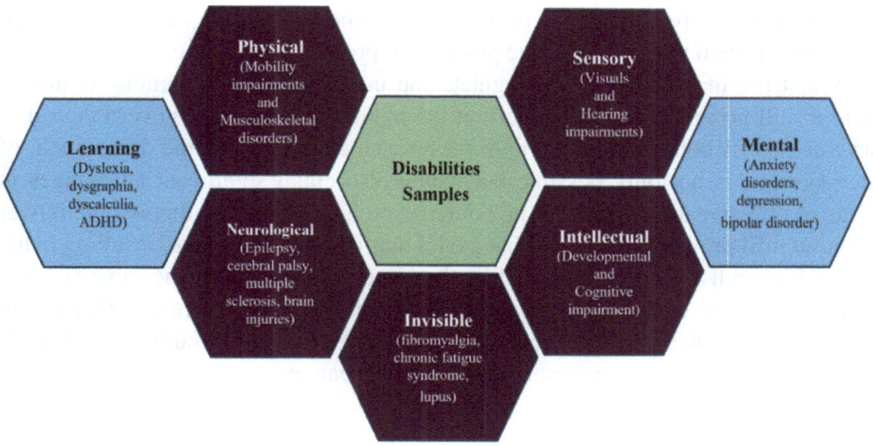

Fig. 2.3 Samples of disabilities

second, and more general set, consists of ethical challenges associated with the wider societal, economic, and ethical implications of advanced AI technologies [5]. These topics are particularly pressing for policymakers, who are currently trying to assess how much and where to invest public funds to foster a rapid and sustained stream of AI research, development, and adoption. And how to regulate or use policy instruments to safely guide the direction that commercial organizations are setting for the promotion of software and information technologies. Common challenges and issues include anthropomorphism, data privacy, security, and the use of AI in privacy invasion, surveillance, dominance in warfare, preventive versus preemptive strikes, AI monopolies, and AI management of autonomous weapons. Responses and proposed guidelines consist of preparation, citizen AI assistance, electronic exports, psychological care innovation for limited academic AI, market group requisitions, autonomous weapon bans, the three laws of AI robotics, AI wages and tax, global management, and agency guidelines for AI research and development of AI [23].

AI and the dual use of AI and other technologies are not new. The potential to exploit the dual use of AI technologies, the risks this entails, and safeguard measures have been anticipated and discussed across society and governments. Challenges may reflect an ethical conflict between national security interests and the potential risk of limiting the freedom of research and exploitation of AI in the civil domain by enacting internationally binding legal agreements. The precautionary principle in policymaking could lead to the adoption of rules that would slow down the use of such technologies, thus weakening their innovative forces. Furthermore, the lack of comprehensive norms or binding regulations aimed at mitigating the undesirable effects of the use of AI-related technologies raises the concern of developers being driven towards quantitative and qualitative unknown limits by the protraction of living with ambiguity rather than taking a short step towards safety [24]. The precautionary principle can sometimes conflict with the enhancement of our life conditions. Many experts think we are playing a very dangerous game on the edge of a precipice, but not many of them would escape the promise of products and profits.

Whilst the majority of states regulate the use of lethal autonomous weapons, there is still no regulatory framework at the international level to govern their use. Several experts believe that we might soon reach a point where the deployment of such systems is not only possible but would also entail significant advantages in terms of effectiveness, precision, and deterrence. Is it ethical to impose a veto over technological progress in the fear that a fatal malfunction may occur, resulting in the loss of human lives? Shouldn't we fear the same in the case that the same systems may help prevent the outbreak of new armed conflicts [25]? Although living with artificial intelligence may soon become a reality of everyday life, culture is not ready yet to accept the daily use of lethal autonomous weapons.

2.5.1 Accountability

As the role of AI becomes more significant in various areas, the question of responsibility is raised. Researchers have proposed three types of AI: one with weak autonomy as an assistant to humans, another with intermediate autonomy as an autonomous machine equivalent to humans, and a third with strong autonomy as a machine superior to humans. Since strong autonomy is still in the theoretical stage, the responsibility of an entity in the novel role of decision-making and the manifestations of its behavioral responses should be derived. When AI fails, responsibility issues lead to a lack of trust and loss of market power, while new regulations and legal prohibitions may damage stakeholders. Accountability in AI requires both internal and external aspects. Internal responsibility involves ensuring that AI acts in a way that aligns with corporate social responsibility policies, regulations, or codes of conduct [26]. These three reduction modes can address the harm of an event caused by the outputs of the AI system.

Explainability/transparency focuses on ensuring that the audience can understand why and how the operation of an AI model used in decision-making was reached. The AI service is human-centric and carried out based on ethical regulations. This requires processes for providing stakeholders with reasons and justifications for AI-based decisions. The explanation technology is based on the AI technology's transparency that shows the AI decision chain and the reasons for its operation. Bias highlights the need for models to be fair in their explanations of the reasons for AI model decisions. The topic of preventing unfair bias is closely related to social responsibility and fairness, including the need to consider social discrimination [27]. This uses AI to analyze and identify social identity groups by means of enhanced learning based on transparency.

2.5.2 Transparency

With the development of AI and its application in daily lives, transparency can hardly be ensured for most people. Because data is the most important factor for AI, and the gathering of most data may be so large and comprehensive that most people cannot access it. Only parts of the data set can be disclosed, thus causing problems. Individuals' privacy would also be at risk due to the huge amount of data required. Another issue is that users do not know how the computer makes decisions and are concerned about computer bugs and failures. The "knowledge gap" results from the fact that most people cannot understand the complexities of many AI systems, especially the reasons those systems generate decisions. However, if an AI system is to fulfill its functions justly, certain aspects of an AI system must be fully transparent, assuming that transparency makes it possible [28]. After all, how can users trust an algorithm whose logic is only understood by a few experts? To address these concerns, development in boosting both the interpretability of the workings of AI systems, which refers to the extent to which AI systems can explain their inner workings to users, and the intelligibility of AI systems that allows the system to be easily understood, should become a priority. For example, it has been suggested that AI systems be programmed in ways that allow the systems to record their decision-making if necessary. This approach is actually borrowed from legal culture and is often used in the highly regulated fields of aviation and nuclear energy. Taken further, a legal requirement might eventually be established that certain minimum levels of transparency, when combined with other measures, are prerequisites for the use of AI systems in a given sector requiring user protections [9].

2.5.3 Fairness

Many AI systems take potentially impactful decisions for humans, such as jail or parole decisions, hiring decisions, credit scoring, or scholarship allocation. These systems can automate and scale decisions, reducing costs and increasing the speed and reliance of the underlying process. However, they may also be detecting or encoding social biases from the data on which the systems have been trained. For instance, it has been shown that a well-known commercial AI system for hiring decisions displayed gender biases. This led the system to penalize resumes containing words such as "woman" or "female" and to favor men for higher-paying technical jobs. The cause might be, at least partially, the existence of biased historical employment data, for instance, showing that many high-level positions have been held by men from a particular social class [25]. Research on fairness in AI has also expanded to include algorithmic fairness when faced with different threat models, a broad set of societal fairness problems, among many other directions. Because the deployment of AI systems in the real world often affects human lives, AI fairness is a timely and important research direction. AI professionals and lawmakers are involved in a lively debate on the possible use of AI ethics and their regulation in the real world. Consequently, AI fairness incorporates social, legal, and ethical rules. Security and privacy provide overarching concerns for AI fairness. Structural AI fairness aims at making all AI systems free from unfairness [29].

2.6 AI in Various Industries

The use of AI and computational algorithms to simulate human work is being leveraged in a number of fields. AI programs can outperform humans in certain cases, which could open the door to significant ramifications in a number of fields, particularly if service industry workers are replaced. These algorithms operate without bias, which could potentially ameliorate certain problems existing in the fields of organizational behavior and human resources [30]. Financial services will have to change tack even more towards these AI algorithms in order to set prices for their services and products. The finance sector, however, has already considered applications of AI in the services they provide. One such application is the use of AI algorithms to predict the stock market, and in healthcare, AI uses include the interpretation of clinical data, pathology results, as well as processing and record-keeping of patient healthcare history [31]. Some applications are in roles that possess life-or-death responsibilities and require 24-h permanence, such as emergency rooms, or can shoulder some of the workload of the overly exhausted professionals, particularly those affected by sleeplessness due to long hours.

2.6.1 Healthcare

Healthcare is a domain where AI is of great interest and much research is underway. It is an area that would greatly benefit from advances in AI, with potential improvements in diagnostics, health monitoring, prescription drug management, and even robot-assisted surgery. International challenges have underlined the importance of this domain. Other primary tasks could also be supported, such as identifying patterns in the development of diseases, personalizing therapy choices in real time, identifying drug contraindications that were not known, and enabling single-person clinical trials. The unique attribute of this domain, compared with many others where AI projects are usually applied, is that the stakes are lives, not simply management, status, or money [32]. This alone makes healthcare AI a special domain, as the use of AI could be the difference between quality of life, independence, and life in general. This issue alone has seemed to slow down research, development, and deployment of powerful AI tools for healthcare. Other important factors might be access to secure data, costs, accuracy rates, and the advances of other areas, such as data mining and widespread autonomous wireless data monitoring systems. Nevertheless, the medical community appears to be finally embracing AI, with many grand challenges being proposed; however, doubts are still present. Success in the healthcare field could give a feeling of confidence to tackle other global changes needed [33]. Benefits from the use of AI in healthcare could be enormous, with the potential for both driving down treatment costs while still providing superior outcomes to those currently received.

2.6.2 *Finance*

Banks and financial institutions are increasingly relying on AI and machine learning to achieve cost savings, make faster and more precise loan decisions, and provide better and safer service. A significant percentage of financial services executives are planning to implement AI technologies in the next 12 months, for reasons such as fraud prevention, risk management, research, algorithmic trading, and customer service, while standalone investment apps use machine learning to optimize investment portfolios. Such AI start-ups go further, offering their customers personalized wealth management and investment services. The challenge for the financial industry, however, is not just to adapt to the developments of technology, but to understand it and use it in accordance with ethical principles [34].

Several financial experts and banks have already begun to implement them in accordance with ethical guidelines. A central bank has developed a governance framework for its AI and machine learning, while a prominent figure in the banking sector has called for a global ethical code to govern the use of AI. Banks have AI and data mining units that develop applications for bank services and products, while a major bank opened a global AI center, representing the bank's priority for advanced technology. AI and machine learning not only can help banks better understand their customers for better personalization of their products and services, but they can also reduce costs by facilitating labor-intensive services like fraud prevention. On the other hand, regulators and boards of financial institutions must remain vigilant, as AI and machine learning also have potential risks that could compromise customer relationships and can lead to ethical issues, as well as an increase in the risk of financial institutions and systemic risk [26].

2.6.3 *Transportation*

When we compare the functioning of large cities in a developed country and an underdeveloped, small or medium country, inequality is visibly evident not only in the city and its structure but also in the technological advances of the transportation systems. New methods, ways, and routines aimed at improving this service have been developed with the arrival and implementation of artificial intelligence, aiming to make it a separate intelligent city with evolving mobility. This smart mobility includes driver assistance systems, driverless transportation systems, and the navigation of self-adapting vehicles, which integrate the production grid [16]. Passenger transportation requires an improved architectural structure, smart strategic management systems, automation, and dynamic ticketing. Transport planning, especially in the developing world, is particularly challenging. But combining the capabilities of multi-agent models with the power of high-performance computing and the availability of massive data generated by human movements optimizes the development and validation process, allowing artificial intelligence tools and automatic learning to predict possible behaviors, facilitating the creation of more advanced strategic planning for the city and enabling better and easier service management by other smart services [13].

2.6.4 Education

A point-bearing postscript. So far, so good, but what is the plan for adjusting educational systems so they can help most people develop the abilities they need to contribute to and possibly share the benefits of advanced computing technology? There are specific challenges learners face and educators must help them overcome are these including learners are not being taught how to ask good questions; that is, questions that are both feasible to answer and worth obtaining answers to because the knowledge they convey will be helpful in making important decisions or solving important problems that learners care about. Another way to phrase this is to say that the "learner-generated" questions being asked are less than totally inspiring [17]. When asked to name the most valuable skills they used at work, a significant percentage of respondents picked "problem solving" over skills like "verbal" or "interpersonal communication," "writing," "research and analysis," or "computer." However, the conventional approach to problem solving in education and in the software that supports it falls far short of equipping students to solve the sorts of complicated, interconnected problems they typically care most about, particularly the important problems they may confront later in life after they graduate, as summarized in Table 2.1.

2.7 The Future of AI

Human-level AIs have been part of science fiction and future-looking conferences for decades, and each new bit of AI progress gets subsumed by sensationalist hype and then assumes its low level of mediocrity once again. AIs have America's political and legal process in a chokehold, but it's mostly hidden from our senses. Despite setbacks or limitations, real improvements could increase the power of these legal AIs, which are too powerful for our own good, with the race condition being AI enhancements rather than AI rights. Although human-level AIs are already much discussed, it is unclear when or what their daily interactions with us will be like. At best, by understanding the motivations of AIs without actually having them, we are

Table 2.1 The double-edged impact of artificial intelligence: sector-wise opportunities and challenges

Sector(s)	AI application	Positive impact	Negative impact	Ethical concern	Mitigation strategy
Healthcare	Disease diagnosis	Improved accuracy and early detection	Potential misdiagnosis due to biased data	Data privacy and bias	Diverse datasets, explainable AI
Education	Personalized learning	Tailored content for individual needs	Over-reliance may reduce human interaction	Data misuse, equity of access	Transparent data use, hybrid models
Finance	Fraud detection	Enhanced security and reduced fraud losses	Discriminatory lending decisions	Algorithmic bias	Regular audits, fairness constraints
Employment	Recruitment algorithms	Faster hiring, skill-based selection	Unintentional bias in hiring	Discrimination	Bias testing, human-in-the-loop review
Transportation	Autonomous vehicles	Reduced accidents, increased efficiency	Safety risks, job losses in driving sector	Liability and accountability	Regulation, human override systems
Social media	Content recommendation	Personalized user experience	Echo chambers, misinformation spread	Manipulation, data exploitation	Algorithm transparency, user control
Law enforcement	Predictive policing	Efficient resource allocation	Racial profiling, privacy invasion	Civil liberties	Ethical frameworks, community oversight
Agriculture	Smart farming	Increased yields, reduced waste	Displacement of small-scale farmers	Tech access inequality	Inclusive innovation policies
Environment	Climate modeling	Better forecasting and disaster planning	Energy consumption of AI models	Sustainability concerns	Green computing, energy-efficient AI
Customer service	Chatbots and virtual agents	24/7 service, cost reduction	Lack of empathy, complex queries unresolved	Job displacement	Upskilling workforce, hybrid services

reduced to that of "pets" rather than that of domesticated slaves. Although AIs with only the current level of legal insight can already do most legal work without human help, we are far from achieving the output of even the very simplest of professional Ais [32]. However, the chaos of our near future is an opportunity for chaotic legal systems to reconfigure themselves into a kind of global legal AI, one that responds to human problems based on their accuracy, focus, and faith in the legal writing process. Since large portions of the legal profession acknowledge the need for regulatable codes of conduct influencing the actions of AI lawyers, companies successfully compete over selling the same legislation that reflects these legal desires.

2.7.1 Trends and Predictions

Tony Tether, the director of the Defense Advanced Research Projects Agency, remarked in 2002, "The role of the pilot has changed over the years, but the importance of having the pilot 'in the loop' for critical events (especially if the automation makes a mistake) has pretty much remained constant. Future automated 'nannies' were most assuredly going to be infinitely more capable than their predecessors. They would be capable of flying a plane to a better conclusion than all but a few pilots in their specific domain. Nevertheless, pilots would always be needed in case of the most fantastically unanticipated circumstances. In fact, the less such events occurred, the greater their unexpected consequences, and correspondingly the less prepared a human 'nanny' would be to deal with the situation [35]. Numerous analyses and simulations using modern methods have borne this out: unpredictable errors by even the most flawless avionics were most likely to produce the most catastrophic failure modes in a pilotless aircraft." As he has retired, he no longer can give us his most candid thoughts. But he has raised a potentially severe problem that is likely to be the Achilles' heel of reaping the value of trustworthy systems. We shall label the phenomenon as Trend #4: Trust without comprehension is likely to create life-threatening situations. It is not only that there is a solution void, but neither the research community nor industry has invested much into it. Our Touch of Trust results involving student populations have concluded that vast, superbly complex A.I.-enabled black boxes may be capable of significantly contributing to a human's well-being [25].

2.7.2 The Role of Government Regulation

Ultimately, the risk of harmful use of artificial intelligence is one that needs to be handled not only by technology companies and computer scientists, but also by society as a whole. Since governments have some sway over the regulation of the mediums through which artificial intelligence is often developed, and in the development and enactment of laws that govern the use of artificial intelligence, they will undoubtedly play a key role in shaping the future development and deployment of artificial intelligence. The biggest risk from the emergence of artificial intelligence is that it will not only be used by humans, but that it will also make decisions that have—or appear to have—a serious impact on humans. Those who are affected by these decisions will certainly want to have some influence over them [29]. People who are affected by a decision are said to have an interest in the outcome of the decision. However, to protect their own interests, they will now have to also protect the interests of the entity that made the decision, since the entity is going to be on the hook if the decision goes wrong. Focusing on the interests of the entity is a new burden on affected parties, since it usually involves new actors that these parties may have no influence over. It represents a serious complication for public governance. At the same time, this burden allocates an important—and potentially life-changing—decision-making role to those entities, thereby introducing them as interesting—and to many, of course, very worrisome—actors on the public stage [36].

2.8 Case Studies of AI Implementation

Our final series of inquiries attempts, with more confident steps, to develop the spectrum of issues that must be explored for an informed consideration. AI does, of course, cover a wide domain of issues and applications, as we have tried to see in conceptual context in the previous chapter. A comprehensive examination of relevant potential issues would take us far afield. All the same, it is important to ground any general speculation about the emergence of any new technology on an understanding of actual problems or issues in an actual setting. We will develop our case studies in a series of inquiries. We naturally start with an examination of problems and potential problems in the financial services industry. With the reformulation of the set of motivations that led to the specific program of credit controls, we see that some have already found such results. But a great deal is still to be understood [31]. AI is so multifaceted that our group found it helpful to ground its inquiry in industry-specific areas where AI technologies have been in use, or at least logically could be expected to be found. We began with the two largest groupings in the original overview: industrial and service activities. Individual industry sectors are of quite variable size and of vastly unequal applied sophistication. We soon seemed to be diving in a fishbowl, in that the number of groups that wanted to focus on the problems of expert systems in the financial services industry always seemed a bit larger than the others. Both groups reported early that there was more to be understood than time greatly permitted [33]. With the step accomplished by these two big general groupings, conservative engineering and the argumentation aspects, five industry-specific case studies were conducted on the application of AI, covering process planning, transportation, insurance, leasing, and pharmaceuticals.

2.8.1 Successful AI Projects

Applying artificial intelligence to an existing business requirement can often create significant value. AI can be used to simulate the actions and activities that a person or team would traditionally undertake. AI applications can encompass many aspects of a role, such as knowledge and decision-making, as well as automation of routine and non-routine manual tasks. AI requires similar foundations to a typical process reengineering program to start. There needs to be a clear and galvanizing reason for undergoing the AI change program. Resistance can be high, but it can be minimized by ensuring a strong reason for change and by taking the workforce along on the AI journey; generally, AI is very adept at providing leaders options to solve business problems they couldn't solve previously and illuminating the issues leaders may not even know they have [34]. An AI change program requires more than technological innovation to be successful. Population of AI applications with real world knowledge and access to applicable data, rigorously defining the outcome measures that will be used to measure and drive success, and a comprehensive adoption and support plan are equal in importance. AI can be used to assess risk and free data assets from time-consuming, error-prone analysis to do more value-added, impactful work. Established approaches exist and are filled with many low-hanging fruit, ripe for plucking. Preparing and augmenting processes beforehand results in the successful deployment and operation of AI applications at scale [17]. However, as with all successful technology deployments, AI adoption doesn't solely rely on technology.

2.8.2 Failures and Lessons Learned

There are a large number of examples of promising applications of AI that failed because the algorithms could not learn effectively from interaction with a complex and changing environment. Several of these failures have led to valuable lessons about the limitations of AI. Many of the early enthusiastic but ultimately unsuccessful projects during the first two "AI winters" of the 1970s and 1980s, when breathless hype about super intelligent machines too often collided with the harsh reality of massively underperforming AI programs and so-called "combinatorial explosion." Among the most high-profile of the early failures of AI is the "shallow-qua" problem in computer vision. This dates back to a much-ballyhooed expert system called "the shallowest generative theory of human perception"—the astonishingly bold and attractively literate construct coined in that paper [14]. The pattern recognition system took a picture as input, divided it into an ungodly number of "ponens boxes," each containing "hypothetical objects," their "descriptive properties," and the "types of parts that they and their properties might reflect."

2.9 Public Perception of AI

Public perception of AI plays a role not only in public policy but also in industry, especially when it comes to adoption. Consistency, transparency, reliability, and safety are key for public trust but are not always easily combined. On one hand, the general public expects AI to be as reliable as a basic calculator. On the other, people are afraid AI will destroy civilization. The issues facing AI are magnified through a mix of public opinion, lack of understanding, and inconsistent news coverage combined with a flood of fake news, not to mention movie fiction depicting AI in a less than favorable light. Furthermore, public policy can differ based on who you are. Offenders lacking privilege are hurt more by AI gone wrong, and there is a different political statement behind outcomes deemed unacceptable for certain parties. Cultural biases and international politics play leading roles in how AI is both perceived and legislated [31]. Immigration officials began using AI to determine whether a traveling couple's marriage is real or fraudulent. Validated with machine learning, the tool drew conclusions from exhibiting behaviors at odds with true love.

2.9.1 Media Representation

The media has a significant influence on shaping the perceptions and attitudes people have towards technologies. Many studies have shown how media representations of various technologies impact public interest, understanding, and opinion towards those technologies. For example, the portrayal of benefits versus risks of synthetic biology can significantly influence public opinion. The same is true for representations of genetics and genetically modified food. The ethical implications of technological innovations that are frequently given the most attention are sometimes quite superficial and wrongheaded [2]. This is often because the ethical debate in the public and in the media is often guided by the use of fictional versions of technology rather than an understanding of what the real technological possibility might be. The way technologies are represented in the media not only shapes public opinion but also affects what politicians and policymakers think and do. Scholars have long observed reciprocal flows of influence between science, the media, and the public. Press coverage isn't just a one-way promotional tool for science and technology [20]. It also shapes the agenda for debate in ways that influence what participants say and how they say it.

2.9.2 Public Awareness and Education

Public awareness, debate, and education in the area of machine learning, artificial intelligence, and their impacts are crucial, especially as there is clearly public interest, currently without public awareness. Beyond the flashy imagery surrounding the field, AI operates in a somewhat obscure and idiosyncratic way. It makes sense to make the citizens more aware of what can be expected from AI, as well as what cannot be expected and, thus, reside in those areas which are subject to human decision and control, or even political decision-making. It is important to create some kind of public awareness about the increasing automation of tasks, including the use of artificial intelligence for interpreting large sets of data. While the idea of a smarter future sounds appealing, it bears the risk of people becoming relaxed and relying on those automated systems too much, even though they should be acting on the results seen, rather than accepting AI's results as the truth. The physical and adaptation mechanics used in various parts of an artificial intelligence system can often be labeled as somewhat obscure, particularly to laymen [25]. To put this in simpler terms, it is often easier to explain and possibly understand how a human learns something new rather than why a certain program solved a particular problem in a certain way, which is particularly difficult for deep and machine learning techniques.

2.10 Collaborative AI

Previous chapters have focused mainly on adversary AI, with good reason. In defending against malign adversaries, necessity requires it. Nevertheless, there are other powerful classes of AI, such as benevolently motivated yet insensitively ignorant 'user utilities maximizers.' In addition, most forms of generative models may be quite easy to collaboratively align—after all, these methods are almost by definition designed to solve a high-dimensional modular addition problem via hierarchical feature engineering. Current work on AI alignment tends to focus on competing abstraction layers and system inequity by module. To the extent that the highest layers are set up in one or more standard ways, the training environment is much like a large-scale online education game. Both researchers and trained entities can take increased power and consequence seriously before and during empowerment [37]. In terms of modeling, the highest layers could then be formatted as supervised policy boards. Various internal posted data can be aggregated and dissent can be tolerated, while the rest are set up to classify either for function composition, various states of the world, conditional meta-level human desires, or advertising side-effect potential to both the user and the community so that unreasonable risks aren't taken. The final classification task can range from collaborative to adversarial, and could be decided by a blend of learned and externally imposed ultimate human time- and instance-varying cost and quality weights. However, as noted, this yields a limited form of corrigibility, not full 'don't kill the operator.' At the highest level is the AI capability. Rather than building an orbital cannon, we could award the system beyond-human-level corrigibility value by default and ask for advice that would help to realign an AI of this capability to speed up things in reality [32].

2.10.1 Human-AI Collaboration

It is precisely because of AI's weaknesses, however, that it makes for such a good complement to humanity's abilities. AI and people have different strengths, and all fields of endeavor are currently limited by the fact that they are not able to meaningfully amplify their capabilities by combining the two. AI can augment the strengths of people and make up for their weaknesses, and vice versa. For example, AI can significantly enhance the soft skills, including cross-cultural communication, of human professionals who use it as part of their work. It can also help people complete an extremely wide range of tasks to a high standard by understanding the specific context and aims of their actions, and be developed to its full empathetic, unblocking, and transformative potential when it is designed to actively involve people with relevant skills, experience, and knowledge in its operations. Through embracing and developing a different role for AI, the future of work possibilities that are currently out of reach for many millions can be realized [26]. AI and people can combine to form a powerful network that is capable of far more sophisticated problem solving than either could on their own, with the corrected and improved ability of decision support and remote collaboration. To properly realize the advantages of AI and people working together requires thoughtful work and continued development of evolving technical and societal systems. This includes collaboration infrastructure, security, privacy, interoperability, robustness, and ethical question management. AI assistance can work at various levels: from providing information, advice, feedback, and negotiation support, saving time and effort, to assuming actual control of a task while input is provided at various levels of granularity [14].

2.10.2 AI in Team Dynamics

Humans typically get upset when others do not follow the rules in a clear situation, such as an intentional foul in sports like soccer. However, it is also pretty clear that if misbehaving is worth the punishment by the referee, it is probably in the team's interest to act unfairly in that situation. As many referees in team sports now receive support from video assistant referees linking AI reasoning about, for example, offside positions, team members will also have to adapt their use of these AIs, which is expected to affect trust and effective team dynamics. More audacious is the idea to include AI and robotics as a team member, whether virtual or physical. Task allocation and foraging behaviors in human groups have been well studied, and we have learned that if all team members are on top of everything, harmful overlaps can reduce overall effectiveness. It will remain to be seen if we can effectively similarly allocate AI-based reasoning processes in mixed teams. Obviously, a key problem is transparency: which reasoning processes can be effectively outsourced to AI, and which reasoning tasks should remain with the team members [31]? The presentation of such AI outcomes has to take into account that team members may have different abilities to understand complex AI-based recommendations.

2.11 Conclusion

Artificial Intelligence represents the promise of a new wave of technology that may multiply output and increase the living standards of people. At the same time, we cannot underestimate the real dislocation from AI. The economic impact of the first and second industrial revolutions is a powerful reminder that it may take time for the transition to a new paradigm of growth and opportunity. And we cannot afford income disparity to explode as is beginning to happen now, when we have so many millions of workers suffering the ravages of technological changes, the financial crisis, and the wrenching changes to the world economic order. The timeframe for adapting to these changes will be crucial. Further, we must not underestimate how complex and intractable these pathways will be. If history is any guide, the industrial revolutions have produced exponential progress and growth, have multiplied people, opportunities, and risks, and ultimately have increased the availability of goods and wealth, advanced equality, and contributed to expanding democracy. But we also know that, meanwhile, in the nineteenth and twentieth centuries, industrial revolutions have increased the wealth of nations and have exacerbated the income inequality within them. They brought about a transfer from the less developed countries to the more advanced ones and blew up a real financial whirlwind; they were responsible for the wars; they gave power and the exploitation of the weakest to the large companies, and the poverty of the subjects.

References

1. Begou N, Vinoy J, Duda A, Korczynski M. Exploring the dark side of AI: advanced phishing attack design and deployment using ChatGPT. In: 2023 IEEE Conference on communications and network security, CNS 2023. 2023.
2. Akter S, Dwivedi YK, Biswas K, Michael K, Bandara RJ, Sajib S. Addressing algorithmic bias in AI-driven customer management. J Glob Inf Manag. 2021;29(6).
3. Chen S, Qiu H, Zhao S, Han Y, He W, Siponen M, et al. When more is less: the other side of artificial intelligence recommendation. J Manag Sci Eng. 2022;7(2).
4. Shafik W. Ethical and legal considerations in artificial intelligence. In: AI-Enabled threat intelligence and cyber risk assessment. 2025;90.
5. Shafik W. Generative adversarial networks: security, privacy, and ethical considerations. In: Generative artificial intelligence (AI) approaches for industrial applications. Springer;2025. p. 93–117.
6. Shafik W. Toward a more ethical future of artificial intelligence and data science. In: The ethical frontier of AI and data analysis [Internet]. IGI Global;2024. p. 362–88. https://doi.org/10.4018/979-8-3693-2964-1.ch022.

7. Shafik W. Science of emotional intelligence. In: Enhancing and predicting digital consumer behavior with AI [Internet]. IGI Global;2024. p. 284–310. https://doi.org/10.4018/979-8-3693-4453-8.ch015.
8. Shafik W. Generative AI for social good and sustainable development. In: generative AI: current trends and applications. Springer;2024. p. 185–217.
9. Ma YM, Dai X, Deng Z. Using machine learning to investigate consumers' emotions: the spillover effect of AI defeating people on consumers' attitudes toward AI companies. Internet Res. 2023.
10. Shafik W. Generative artificial intelligence for social good and sustainable development. In: Generative AI: Disruptive technologies for innovative applications. 2025;153–85.
11. Hidayatullah AF, Kalinaki K, Aslam MM, Zakari RY, Shafik W. Fine-tuning BERT-based models for negative content identification on Indonesian tweets. In: ICITDA 2023—proceedings of the 2023 8th international conference on information technology and digital applications. 2023.
12. Wirtz BW, Weyerer JC, Sturm BJ. The dark sides of artificial intelligence: an integrated AI governance framework for public administration. Int J Public Adm. 2020;43(9).
13. Wiederhold BK. The dark side of the digital age: how to address cyberbullying among adolescents. Cyberpsychol Behav Soc Netw. 2024;27(3).
14. Mikalef P, Conboy K, Lundström JE, Popovič A. Thinking responsibly about responsible AI and 'the dark side' of AI. Eur J Inf Syst. 2022;31
15. Papagiannidis E, Mikalef P, Conboy K, Van de Wetering R. Uncovering the dark side of AI-based decision-making: a case study in a B2B context. Ind Mark Manag. 2023;115.
16. Cheng X, Lin X, Shen XL, Zarifis A, Mou J. The dark sides of AI. Electronic Markets. 2022;32.
17. Zhou Y, Wang L, Chen W. The dark side of AI-enabled HRM on employees based on AI algorithmic features. J Organ Chang Manag. 2023;36(7).
18. Rana NP, Chatterjee S, Dwivedi YK, Akter S. Understanding dark side of artificial intelligence (AI) integrated business analytics: assessing firm's operational inefficiency and competitiveness. Eur J Inf Syst. 2022;31(3).
19. Satornino CB, Grewal D, Guha A, Schweiger EB, Goodstein RC. The perks and perils of artificial intelligence use in lateral exchange markets. J Bus Res. 2023;158.
20. Lv L, Huang M. Can personalized recommendations in charity advertising boost donation? The role of perceived autonomy. J Advert. 2024;53(1).
21. Pearson A. Refrigeration applications column the dark side of AI. ASHRAE J. 2023;65.
22. Xiao L, Shen XL, Cheng X. Introduction to the HICSS minitrack "the dark sides of AI". In: Proceedings of the annual Hawaii international conference on system sciences. 2022;2022-January
23. Shafik W. Deep learning impacts in the field of artificial intelligence. In: Deep learning concepts in operations research [Internet]. New York: Auerbach Publications;2024. p. 9–26. https://doi.org/10.1201/9781003433309-2.
24. Grundner L, Neuhofer B. The bright and dark sides of artificial intelligence: a futures perspective on tourist destination experiences. J Destin Mark Manag. 2021;19.
25. Gligor DM, Pillai KG, Golgeci I. Theorizing the dark side of business-to-business relationships in the era of AI, big data, and blockchain. J Bus Res. 2021;133.
26. Cao L, Chen C, Dong X, Wang M, Qin X. The dark side of AI identity: Investigating when and why AI identity entitles unethical behavior. Comput Human Behav. 2023;143.
27. Sun Y, Li S, Yu L. The dark sides of AI personal assistant: effects of service failure on user continuance intention. Electron Mark. 2022;32(1).
28. Shafik W, Singh R, Kumar V. Artificial intelligence transparency and explainability in sustainable healthcare. In: Transforming healthcare sector through artificial intelligence and environmental sustainability. Springer;2025. p. 165–91.
29. Castillo D, Canhoto AI, Said E. The dark side of AI-powered service interactions: exploring the process of co-destruction from the customer perspective. Serv Ind J. 2021;41(13–14).
30. Shafik W. Security, 15 privacy, and trust in Fintech. FinTech and financial inclusion: leveraging digital finance for economic empowerment and sustainable growth. 2025;216.

31. Barman D, Guo Z, Conlan O. The dark side of language models: exploring the potential of llms in multimedia disinformation generation and dissemination. Mach Learn Appl. 2024;16.
32. Ojha NK, Pandita A, Ramkumar J. Cyber security challenges and dark side of AI. 2024.
33. Pantano E, Marikyan D, Papagiannidis S. The dark side of artificial intelligence for industrial marketing management: threats and risks of AI adoption. Ind Mark Manag. 2024;116.
34. Belanche D, Belk RW, Casaló L V., Flavián C. The dark side of artificial intelligence in services. Serv Ind J. 2024;44(3–4).
35. Kumar G, Singh G, Bhatanagar V, Jyoti K. Scary dark side of artificial intelligence: a perilous contrivance to mankind. Hum Soc Sci Rev. 2019;7(5).
36. Alawida M, Abu Shawar B, Abiodun OI, Mehmood A, Omolara AE, Al Hwaitat AK. Unveiling the dark side of ChatGPT: exploring cyberattacks and enhancing user awareness. Information (Switzerland). 2024;15(1).
37. Xing Y, Yu L, Zhang JZ, Zheng LJ. Uncovering the dark side of artificial intelligence in electronic markets: a systematic literature review. J Organ End User Comput. 2023;35(1).

Chapter 3
Ethical Dilemmas in AI: Navigating the Moral Minefield

3.1 Introduction

In thinking about ethical quandaries related to artificial intelligence, it is useful to examine the origins of the field to understand where some of its particular challenges have arisen from. The idea of artificial intelligence has existed in one form or another for quite some time: stories about humans creating "artificial slaves" can be found in ancient literature across the globe, while the Czech author famously coined the term "robot" in 1920. However, the modern field of AI was truly born in 1956 at a conference, the brainchild of a mathematician who thought deeply about the social implications of automating the decision-making processes traditionally part of the human condition [1]. The use of machines to solve symbolic problems using algorithms was a radical departure from previous engineering work, where problems were either simple and required only logical elements to solve, or complex deterministic problems addressed through brute force calculatory strategies.

Early optimism, money, and high public interest on the part of both the private and federal sectors spurred developments in AI throughout the 1950s and 1960s. Both heavy machinery and deeply algorithmic systems were seen as having the potential to "liberate" human beings from the toil of regular work. Mitigating ethical and policy challenges was certainly recognized [2]. A notable work highlights the potential social implications of automation and sophisticated, large-scale mechanical systems that rely on abstract non-symbolic reasoning. In parallel, two of the handful of inventors of early machines capable of learning new knowledge emphasized the potential benefits of these machines: "A low-grade make-it-yourself intelligent machine could engage more of the intense attention and love of its companions over a long period than the best animal ever could" [3].

W. Shafik, *The Dark Side of AI*, https://doi.org/10.1007/978-3-032-09130-7_3

3.2 Key Ethical Theories in AI

Deontological ethics suggest that the right thing to do is to follow your duties and obligations. It also emphasizes following rules and procedures as well as obeying orders from people in authority. It also concerns the intentions behind one's actions that determine whether actions are good or bad. On the other hand, consequentialism is associated with the outcomes or consequences of actions. The right move is the one that leads to a better outcome. The better outcome in this sense usually means the one that results in the greatest happiness for the greatest number of people. Utilitarianism is a branch of consequentialism. It emphasizes social welfare maximization [4]. According to utilitarian ethics, the number of people affected does not matter; what matters is the net change in the welfare of individuals affected by an action or a policy. Rule utilitarianism ought to obey utilitarian rules in providing the greatest net welfare and should not take action on an ad-hoc basis. Finally, virtue theory is concerned with the nature of the moral agent, what sort of person someone should become, rather than with the ad hoc judgments that deontological ethics and consequentialism can generate. Vices and virtues focus on the person displaying them more than on the nature and consequences of actions taken [5]. According to virtue ethics, moral behavior is similar to an art in which the moral agent displays virtues like honesty, truthfulness, courage, and so on. It should be kept in mind that no action is intrinsically right or wrong. Rather, actions are judged by the virtues that these actions demonstrate. Theory-based ethics is another approach taken to solve ethical dilemmas by incorporating moral rules into the decision-making process. Theory-based ethics is binding by the fact that rules are rules. Theory-based ethics is similar to deontological ethics in that some rules cannot be broken. Although rule violations in some cases may seem morally acceptable, this does not mean we should have this possibility [6]. It is also like a company HR policy that conducts the termination of employees when company rules are violated.

3.2.1 Utilitarianism

Utilitarianism is the ethical theory that holds that the best action is the one that maximizes the general good. This theory often applies to groups of sentient beings or simply to sentient beings, where the latter group refers to living organisms that are capable of experiencing pain and/or pleasure. The seminal illustrations of utilitarianism in history are connected with the names of philosophers. They held that the ethical norm of any given situation should be defined according to the principle of utility, which equates goodness with pleasure and absence of evil with the lack of pain. To define the degree of utility associated with a given action, the philosophers counted the amount of pleasure and pain it caused, taking into consideration its intensity, duration, sureness, and/or propinquity [7]. In the opinion of the utilitarian philosophers, correctly rated moral values should support not just an individual interest but also a collective interest, concentrating on the community as a whole.

3.2.2 Deontological Ethics

Deontological views, the stance that non-consequential factors can, in certain instances, justify actions based on their following a set of moral rules, is the strongest philosophical motivation for a societally robust AI safety mechanism. Specifically, because humanity either does not have access to all the required ethical knowledge to predict the moral implications of potentially dangerous AI applications, or we are unwilling to compromise our foundational moral standing by pursuing dangerous AI research, we must enforce research boundaries. A necessity or desire for societal safety requires us to restrict freedom. This could benefit other people or it could bring benefits to the person who is bound by the restriction. There is a very strong reason why we should choose whether to give away our concerns for safety in this way [8]. If humans cannot pursue a science without restrictions designed to decrease its dangers, then scientists and policymakers have a duty to enforce safety regulations because they could compromise the moral standing of humankind.

We hear you scream as you flash back to those philosophy lectures you nearly slept through at university—deontology? Consequentialism? What's this got to do with AI safety? Sure, in AI we often think in black and white terms about the safety of our systems (they are either safe or they are not), but that's because we are engineers; our language is the language of measurable outcomes, of equations with an unambiguous result. But switch to the language of ethics, your freedom to act without the consequences of the actions to dictate morality, and you move from a comfortable world of defining probability distributions over outcomes to one where the use of the word "should" is contentious [9]. Should you do X, or should you not do X, and how would you feel if I defined a situation where following your own rules led to a worse outcome for society than doing what was not intuitively wrong but was contrary to your usual ethical obligations? The point being that we are ill-equipped to take final value-based decisions without considering the implications of those actions, and ethical considerations can (and should) be an informative part of our decisions. The world of AI has always seemed immune to these sorts of moral and ethical dilemmas, but that is changing: we now know that the impacts of some AI policies are so great that they have the potential to reduce human ethical principles to a set of trade-offs with moral machines [10].

3.2.3 Virtue Ethics

A third major approach in moral philosophy that is relevant to addressing the crisis in AI and to designing human values into AI is virtue ethics. This is a long tradition that starts with Aristotle and continues to be a major perspective in contemporary moral philosophy. Virtue ethics is fundamentally concerned with the question of the nature of the good life and how it can be achieved. It suggests that to lead a good life, it is necessary to develop traits, known as virtues, that allow a person to flourish. Virtue ethicists are therefore concerned with the kind of person a human should be and the self-cultivation practices necessary in order to become that kind of person [11]. Moral philosophy is of interest from this perspective primarily because it provides basic rules of conduct that can help one to develop the virtues. This contrasts with consequentialism and deontology, which typically assume that a person should be able to recognize and act upon moral actions without necessarily being a good person. From a virtue ethics perspective, moral rules are best not thought of as following algorithmic rules or principles, but as developing virtues that lead their possessor to automatically do the right thing because it is in one's nature to do so. When developing virtues, it is necessary to be a member of a community; these virtues are not developed alone but with others within the community [12]. Indeed, it is only by being in a practicing community that one can develop these virtuous character traits.

3.3 AI and Decision-Making

Artificial intelligence (AI) and the tools through which AI operates are tools to assist and guide the decision-making process and other technology support, as demonstrated in Fig. 3.1. However, AI can upset some of the established paradigms, constituting a point of change in critical contexts. In particular, three consequences arise that deserve analysis. First, whereas decision-making can be frustrated by serious limitations, AI makes it an instrument dependent on the quality of the preexisting rules and the available data. Second, AI is in a position to reveal prejudices, stereotypes, or analog forms of discrimination [13]. At the same time, the risk of partial or faulty use must be recognized for the core capacities of AI and the phenomena of decision-making that AI serves, including ensuring the compliance of the decision model through methods of procedural due process related to the right to an algorithmically transparent administration. In automated procedures and decision-making, the selection of the AI model is crucial to the quality of the final decision; pre-established rules shape the final choice. Yet, AI is trained in the modus operandi of the model and in the methodologies through which it operates. The problem, therefore, concerns the quality of the pre-established rules and the quality of the training data. Where the AI lacks performance due to inefficiencies or faults of the model, or insufficiencies or faults in the training data, the final result is neither valid nor efficient despite the capabilities activated [1].

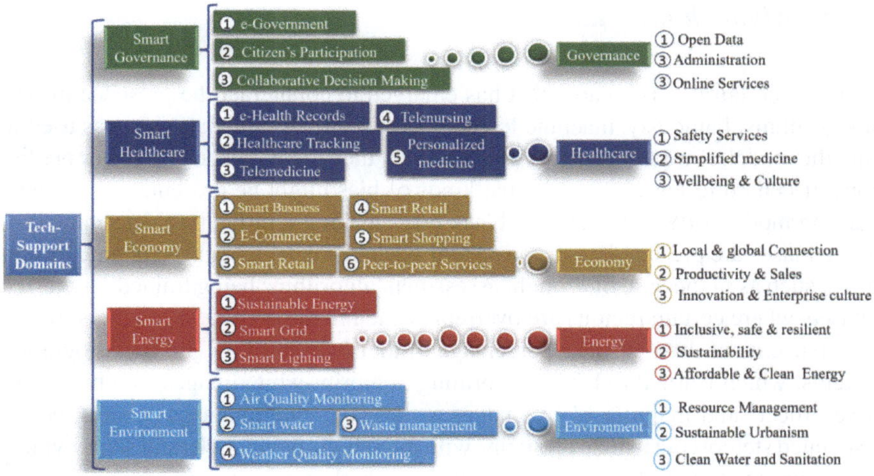

Fig. 3.1 Technology support domains

3.3.1 Autonomous Systems

Another focus of attention in the debate on ethical issues and artificial intelligence has been on the ethical design and use of so-called autonomous systems. The concept of an autonomous system refers to a technology that is not just capable of automated execution of tasks, but also of independent decision-making and action. The emergence of machines that combine computational skills and moral code also stirs the waters of both the AI field and discussions about the future of work. However, the idea that machines should make moral decisions has aroused controversy. For example, should a car be prepared to run over an elderly person or two children jumping in front of the vehicle? This kind of research tries to address this dilemma by predicting which part of the population would opt for each choice in various situations [14]. This is another level of abstraction from the concept of autonomous vehicles but still a pending issue, according to ethical principles.

The prospect of delegating significant moral decisions to machines has led to arguments that delegating responsibility for establishing criteria for choosing between good and evil, life and death, right and wrong to autonomous systems might challenge the human condition itself because our moral capacity is central to the notion of a "moral subject." That is, living beings are ethically responsible only to the extent that they are capable of making moral choices. Moral agency is a quality restricted to those who are capable of autonomous functioning. This relation between moral capacity and moral agency raises discussion not only regarding the delegation of moral responsibilities to machines but also regarding human moral education [15]. If machines cannot become moral agents, then programmed guidance is no substitute for humans taking on the moral duty to use the machines in a way that is consistent with sound ethical principles.

3.3.2 *Algorithmic Bias*

Algorithmic fairness is an area that has emerged to counteract the presence of bias in algorithms. Famously, machine learning mainly utilizes the data that was used to train the model. If we train on data that is biased, then there will be bias in our predictions. It is hard to know a priori what kinds of bias might be present, and machine learning models often operate as a 'black box,' making it difficult to determine if or where biases are present. Biases may occur when the data being used is unrepresentative, such as in the case of drought assessment algorithms being trained on skewed datasets where certain regions are overrepresented to the detriment of others, which experience more droughts as a percentage of their weather [16]. Image recognition systems, which learned to label conforming, majority-white images as 'chief executive officer' and majority-black images as 'inmate,' and language models, which associated stereotypical male pronouns with more high-paying jobs and stereotypical female pronouns with lower pay and domestic work, were shown to be bringing in biased assumptions from their training data [17].

There is concern that these mischaracterizations contribute to algorithmically reinforced oppression, so addressing these issues of bias is quite relevant to the future of AI in the philosophical sense. This concern stretches to the larger fear that technology is very good at maintaining a status quo, and in a world plagued by social and economic inequality, this is making people with low incomes poorer. Bias is not good for business. The loss of a major AI system turned into an embarrassment after it emerged that experts found baffling treatments that AI had recommended, so those biases need to be ironed out if AI is to be welcomed into hospitals or boardrooms with open arms. Researchers working to increase the fairness of AI systems are doing essential work in the fight against oppressive uses of AI; it should be celebrated. It is also beneficial for those tasked with implementing fairness considerations that there is now legislation to back up their internal efforts [18]. The legislation is powerful not only because a major economy is enforcing it, but also because the sponsors represent a range of political positions and a range of industries. With this diversity present, there are hopefully low barriers to other nations following suit.

3.4 Privacy Concerns in AI

As we start seeing machine learning solutions applied to a wide range of domains, we have seen a wide range of privacy incidents arise as well. Some of these stem from naive or simplistic applications of machine learning technology to personal data, without taking into account the amount of re-identification risk in the data that is sometimes present. Technologies like machine learning are also particularly suited for surveillance and mass surveillance. Whether it is to gain microtargeting profiles, automate censorship, or track people based on their ethnic background, it seems as if sophisticated technology-driven privacy violations are increasingly on the rise [19].

In terms of criteria, the risks for privacy involve being surveilled, being banned from private and intransparent automated decisions, and having one's private life exploited. While population surveillance could often be painted as a democratic issue rather than an ethical issue, numerous other questions apply directly to the individual or groups of individuals. For example, those who are already marginalized run a much higher risk of being exploited by connecting behaviorally targeting data from private and public sources. It is easy to identify when someone is about to lose their job or to identify a dissident [20].

3.4.1 Data Collection Practices

Ethical considerations begin at the earliest stage of the AI project life cycle, during data collection. The key question is: what data is collected and used to train and test the AI system, and how is it collected? These considerations often involve entirely different ethical principles from those debated in media headlines and associated advocacy work. Practitioners specializing in these systems often make value judgments and express their sense of practicality and experience by actively deciding not to collect certain potentially useful data due to the risk that certain groups may be unfairly affected by those design decisions later in the system's life cycle. This aspect of AI ethics and AI fact-finding is important and includes both systems used for long-term AI system designs as well as short-term AI forensic purposes, such as distinguishing artificial from live users [21]. The selection of the data that will be collected thus raises a long list of ethical considerations that are often unrecognized outside the design and development community.

3.4.2 User Consent

Companies may say that they're gaining meaningful consent—and sure, clickwrap is simple—nevertheless, it does not change anything. One of the most integral reasons for a consent process is that it signals that a person's sovereignty and autonomy are being recognized. The greater good for society is to have confidence in how advanced technologies are being utilized and their amplification on human decision-making, and how meaningful the supposed consent is required to process the data through numerous other data processing steps. Even though there are some moving parts, meaningful user consent is still fundamental in several types of AI use cases and data processing operations. If there is no honest consent, the organization gets a massive picture of a society that doesn't trust powerful organizations in the digital age [22]. Your ability to draw AI conclusions will be predicated on the volume and quality of training data available to you. Consent providers to everybody are on notice about their obligations. A data subject must be provided with customized access to policies, terms of service, and other essential information, and engage with sections of privacy documents before the data subject's consent is presumed [23]. These rules apply no matter the privacy and data processing status or their considered consent obligations. No more brushing aside privacy laws that all of a sudden they suddenly realize they were in breach of. If only these powerful tech companies would realize that maybe people are mad because they should be. Maybe it is finally time to pivot away from the growing distrust of personal data exploitation and start collecting and processing information, given that people are increasingly concerned about these personal privacy rights, working with the consent that is received [6]. This is usually influenced by a number of factors, as illustrated in Fig. 3.2.

Fig. 3.2 Factors amplifying the dark side of artificial intelligence

3.5 Accountability in AI Systems

When it comes to civil litigation, AI technology is only at the start of the legal Olympic high jump. As AI is not designed to deceive, in most cases, it is still its human creator who will be assigned responsibility under criminal law if something goes wrong. Machines are not capable of being morally culpable in the same way as an individual who has acted with the intention of committing a crime or in circumstances where the criminal provision calls for punishment without intent. However, with the increasing use of algorithms, the time is fast approaching when the behavior of machines will not only need to be classified as programmed, but also subject to a specific risk assessment regarding any civil and criminal liability [24].

Suppose an autonomous system breaches the duty to fulfill guarantees, resulting in damage. In that case, the manufacturer will be liable to compensate for the damage merely in accordance with the Product Liability Act, especially during the time of the pandemic [25]. For other cases in which no person has been killed, and their spouse or relative shipped to some far-off location, or where family, legal protection, data regarding freedom, or personal protection rights could have been affected, no specific provisions of the law currently provide details on any liability for the breach of a guarantee. Related to autonomous systems, the law on liability for defective products must clarify which product defect reduction should apply [26]. This is closely linked to the nature of each product and the reasonable expectations of the buyer or user, resulting from the public product designation, manufacturers, distributors, or service providers, including advertisements.

3.5.1 Responsibility for AI Actions

Even with these drawbacks, it is clear that assigning blame remains a cornerstone of human societies: capable guilty actors are punished and incapable innocents are absolved. As AI becomes capable of finding the correct balance between the undeniable incentives of preventing repetitive errors and unfairly punishing machine intelligence, fear of the former overflowing into precautionary over-caution is unwarranted since responsibility assignment is an important consideration for any AI developer. An approach that upholds the concept of responsibility for AI systems is another important principle related to intelligence attribution. According to the principle of 'proportionate responsibility,' the attribution of responsibility across hierarchies should be roughly based on what each individual is capable of affecting [11]. Accordingly, the larger portion of intelligence should have a greater share of the responsibility attributed to the system itself.

3.5.2 Legal Implications

The fundamental ethical and economic questions require the technical design of AI systems to adhere to a set of principles that are not always straightforward to implement, such as transparency, diversity, non-maleficence, and the right to systematic redress. The combination of ethical inferences derived from the technology design with well-established regulations in various domains raises, in addition to technical issues around the AI capabilities, fundamental questions about the legal framework [27]. Access to intelligence, the use of AI to make decisions, and their societal impact must necessarily be grounded in the rule of law for it to be fair and apply to everyone. As AI becomes a ubiquitous presence in our environment, it is essential to understand AI as not only a subject of regulation but also a tool to improve enforcement and compliance, potentially [14]. By participating in the creation of new laws, AI can help address the challenges posed by technology and inform the formulation of artificial intelligence policies that are not overly prescriptive and risk stifling innovation and scientific progress. Moreover, access to intelligence with objective criteria and redress mechanisms is instrumental to respect the rule of law in countries without well-established legal systems, helping settle legal uncertainties around individual and collective responsibilities. The law governs not only the use of AI; it also limits how this technology can be designed. Suggesting new laws, but also conceiving how AI-enabled policies can enforce these laws, is the focus of this section [28].

3.5.3 The Role of Transparency

The question now arises of why and in what respects companies must enable transparency in the use of AI for job assessment, among many other possible AI applications. We hold that "Accuracy, Fairness, Privacy, and Bias (also sometimes referred to as Interpretability) are ethical considerations that are applicable to any system that uses AI"; then we add the specific case of "job assessment" as an exemplar in order to discuss the nature of these ethical considerations. Taxonomy in hand, some of these considerations will then be discussed at a high level. The demand for transparency results from the need for informed consent and the demand for accountability. Applied to the specifics of job assessment, individuals must know what AI systems know about them, how the system uses this information, and whether these uses are justified. Individuals must also be protected against negative outcomes that result from the use of the system [19]. Such protections are almost impossible to guarantee without a level of transparency that, at minimum, generates confidence in users that outputs and decisions can be trusted.

An automated video interview system analyzes both verbal and nonverbal feedback. It can predict personality and communication skills. There is "a lack of transparency about how algorithms and that information benefits or hurts the individuals they analyze." There is also the potential for algorithms to make interviewer bias worse, because employers might rely on AI to weed out candidates early in the process, hiring a less diverse group of candidates for in-person interviews. The beginnings of the use of AI in job assessment will necessarily lay the groundwork for future decisions about the ethical implications of AI [28]. Although these technologies are in their relative infancies, and current uses focus on specific (although significant) applications, individuals must be protected from the potential harms caused by opaque systems. The first uses established effective standards and toolkits with which small and medium-sized businesses, many of whom have not previously used AI, can assess and implement these new technologies. If those new technologies evolve to be complex, the requirements for these standards and toolkits will only increase. These difficulties (and the implicit needs) mean that small and medium-sized business users may be more affected by unfair and biased AI [11]. Thus, the use of job assessment technologies is where the need and requirement for transparency is most acutely felt.

3.5.4 Explainable AI

Explainability is often cited as a requirement in machine decision-making. Usually briefly termed 'transparency', the desire for explanations denotes a much broader, deeper societal aspiration than mere technical understanding. Possessing a suspicious nature that bears with it the desire to question, rationalize, and comprehend not only the machine but also the sources of the machine, people would appreciate discourse offered in their language. However, at its most pragmatic, the requirement for explanations is most often encountered in situations of decision support or decision augmentation, where the end goal is to guide human decision making in delivering the 'right' response. In other, more utilitarian settings, achieving socially desirable outcomes likely takes precedence over the requirement to explain [29]. Ethical questions arising from transparency fall broadly into those associated with disclosure adequate to achieve trust in decision making and its capability for challenge, including the veracity of the justification, and those relating to the form and natural language expression of that justification. Critically, the explication of explainability is taken beyond a purely technical understanding and is set within the context of societal expectations. Pose this central question: How much transparency in AI-based decision-making practice is optimal in order to maximize trust without overburdening the explanation process and potentially confounding the human [3]?

3.5.5 Trust in AI Systems

Several recent studies have considered the notion of trust in AI, exploring how trust can be fostered, the key factors that affect it, and the potential consequences of trust in AI. However, we are not convinced that trust should be the end goal, particularly in relation to ethical decisions. Indeed, the idea that users should trust AI is more than a little puzzling and potentially dangerous. In general, trust has been associated with machine learning from the earliest days; it covered a wide range of situations, notably including categories [30]. However, the impact of the use of categories by recommendation algorithms raised the trust issue straight away; users' reactions and the context in which they took place highlighted the ethical problems surrounding automatic categorization. The effect of trust was considered in more depth by another group of researchers who studied how recommendation lists influenced the level of trust. The results show how users made use of such lists, as they suggest that appropriately balanced recommendation lists generate a higher level of trust in the recommendation algorithm and, at the same time, can prevent information overload. Hierarchical multi-label classification has been effectively reduced to a series of classification problems where classifiers recognize new categories, providing a constant stream of confirmations to the algorithm manager [31].

3.6 Impact of AI on Employment

There are polarized views on the job displacement effect of AI. Some prominent economists argue that AI will displace many workers and polarize workers into those with in-demand skills who earn good salaries and others who work in jobs with stagnant or declining wages. Others call such claims realistic and argue that, especially in less affluent areas, large-scale changes to how people earn a living will be necessary. On the other side, it is argued that AI may only be the latest technology putting stress on the established hosts of work, just as the industrial revolution did. However, it is noted that those societal stresses could still produce wider inequalities in living standards within different occupations [32]. Despite the broad employment growth historically experienced in the early nineteenth century, specific groups of workers did lose their jobs, it is conceded. Furthermore, existing AI tools like credit scoring algorithms and AI-powered surveillance tools are tilting the balance of power even further towards employers. They are already helping employers control their workforce by disciplining and punishing workers for their undesirable behaviors in the surveillance capitalism age. Surveys of HR professionals suggest that with AI, employers will be able to increasingly enforce the in-demand skills and status of workers by identifying union supporters and political dissidents, or monitoring, predicting, and managing worker resistance. AI may bring new types of technologies that destroy, diminish, and demean the dignity of non-elite workers [11].

3.6.1 Job Displacement

Job Displacement—More than 50 percent of the workers in the United States are classified as "knowledge workers" who do not actually make or deliver a product.

They include accountants, analysts, brokers, teachers, executives, journalists, and many other professions that depend upon rapidly advancing information technology. To the extent that new organizations bring AI into the decision-making arenas of these functions, job changes will or may accompany these changes. It may be that the nature of the new jobs created by AI will not only bring higher pay but may also be termed "fun jobs". Or, in the universities, a return to the Socratic method of learning. However, other kinds of jobs may just be plain lost along the way, and that situation could create quite a bit of social instability [28]. Robotic machines will also displace large numbers of "problem" workers. This will happen in both Russia and what used to be Eastern Europe, as well as in other countries. It is a significant tragedy of modern society that several prominent thinkers have observed that the destruction of groups of individuals can lead to a kind of internal respect that can be quite addictive and not immediately blameworthy. With the power and flexibility of the next generation of AI programs, the destruction may be even more personalized as a practical matter. In the U.S., some retail clerks now make weekly hourly salaries multiplied by the number and quality of their sales [19]. The automatic teller machine has already increased the productivity of bank tellers by about 300%, and robots doing sales will increase future productivity even more.

3.6.2 New Job Creation

New employment opportunities are expected to arise from the automation of certain previously unthinkably complex tasks, particularly in the manufacturing of AI-compatible user-end products such as robots, drones, smart soles, intelligent machines, and navigational software that can make everything function as designed. Other jobs that are expected to be created on account of the AI revolution include trainers and interpreters of intelligent reasoning, machine learning engineers, regulatory specialists responsible for programming compliance, machine biodiversity officers, and various forms of machine-human integration service handlers, including for older people and for emotionally disabled children. As expected, the fast-paced revolution in smart machines, AI, and factory automation is also expected to create several new occupations [12]. The need to lessen investments and the time required to learn and implement workable processes capable of accommodating smart machines will also result in a simple increase in the number of traditional robotics, 3D modeling gatherings, manufacturing employees, and welding workers. The need to meet the demands for products and weed out technical problems and inefficiencies inherently attendant upon modern machines is expected to produce several other employment subtypes and job titles where the genius of man combines with that of the machine. These will range from robot counselors and robot teachers to arrangement alignment experts and robot reserve authorities. The new robotics will, of course, call for adjustments in expanded curriculum at educational institutions as well as in the job class structure to support new levels of skills and assurance of future support [33].

3.7 AI in Surveillance

AI, essentially by enhancing the informational advantage of the powerful, makes social inequality worse. Nowhere is this so true as in its application to surveillance. Powerful states and their secret police agencies are using computer talent to perfect the model of the surveillance state. If there was any hope that the more liberal democratic societies of the West might try to resist this trend as an infringement on the traditional liberties of free societies, it showed that legal prohibitions, even of relatively extreme violations of privacy, were able to be set aside. Today there are very few alternatives to the use of such invasive techniques on the grounds that in a democracy we might have a right to know that danger to society is lurking in our midst, but we also have the right to demand that anyone who uses such powers of surveillance should do so only in a clear framework with an absolute minimum of oversight [34]. The misuse of the information that such techniques collect is evident enough, and their potential for misuse is truly horrifying. But arising from the nature of the techniques is a more insidious problem: the creation of a 'suspicion society' in which everyone has something to hide, everyone is a potential wrongdoer and the omniscient state keeps watch on all.

3.7.1 Ethical Implications

The phrase 'ethics in AI' immediately signals the role of AI as an end in itself, rather than as a means to achieve specific goals. The overriding concern in this context is the belief that computers will begin to compete with humans for ethical rights and responsibilities, and that reasons will be found to attribute moral worth to AI systems. An extreme scenario would endow AI systems with human traits and a human state, which would then give them moral rights. A second argument surrounding ethics is not that AI will compete with humans for ethical status, but that advances in AI have the potential to increase the number, diversity, and complexity of human–computer interactions that have ethical significance. Such systems might make decisions, take actions, or have effects that have ethical consequences for humans, as well as other computers [35]. Anyone who believes that the unique ethical roles and responsibilities humans have for AI systems, therefore, amalgamates the two fundamental tenets of the AI discipline. This fusion carries with it significant practical implications that, at present, stretch conventional ethical concepts, principles, and methods to breaking point. It is now common to find analysis of the ethical and philosophical implications of AI and the broad policy and regulatory frameworks that give ethical concept issues academic weight. These attempts are to provide a cohesive ethical framework to underpin the work of the wider AI scientific community [36]. The new assumption of ethical principles would be interactively engaging, and this would help with this process. The remainder of this section is devoted to doing just that.

3.7.2 *Public Safety versus Privacy*

If fully autonomous AI tools are certified to take on a broader context of decision-making duties, their impairments need to be mapped out and addressed. However, such diminutions are rarely clear and calculable. It can well be that to meet its obligation of due care adequately or to be fair towards the AI developer, the public interest should come first, even if that requires significant impairments of the AI system's capabilities. Not only could this lower the dereliction hazard, but it should incentivize all relevant stakeholders to work on addressing the problems counted under Sect. 9.1. However, the possibility of any imposed impairments remains at best marginal, also due to the fact that even if granted, technical agents would have insurmountable difficulties in satisfactorily addressing and evaluating them [34]. Hither, safety has automatically the higher, more general legitimacy (over privacy, a value that oftentimes can mean a catastrophe averted, not a particular good delivered as a result of a decision, and as such might seem less important to the AI tool). In other cases, when the value identities might change (in the sense that access to personal data can be seen both as guaranteeing social system inclusiveness and stability as well as being an abusive overreach of government security measures), the question then is whether politicians, the citizens, or the AI tools could successfully recognize and differentiate between them [11]. If they can, should AI tools be programmed to preemptively respect the outcome and decide in privacy's favor (rather than at great social cost), or should they rather help only uniformly execute the norm, regardless of the significant social disruption it might cause?

3.8 Bias and Discrimination in AI

The manifest discriminatory consequences when algorithms are deployed in opaque ways have raised alarm for concerns ranging from hidden biases to what some label as the new 'digital apartheid.' Definitions, accountable to notions of justice, of ethical and socially responsible behavior, coupled with regulatory frameworks that require transparency for complex and proprietary algorithms, although needed, remain insufficient as they struggle to deal with the inherent uncertainties related to and efforts of private vendors to muddle the use of protected proprietary information as being related to the risk associated with being 'claimed' to be discriminatory. Still worse, black-box algorithms can often make spotlighting biased decisions nearly impossible [19]. Technical framings generally ignore the inherent complexities of understanding and making trade-offs in profoundly messy social domains. Embodied expertise in directly engaging and valuing the input of those who have historically been marginalized is necessary in training data circumstances. Varied implementation stories will guide us on our paths to more robust and fair systems. We can, from the very beginning of the engineering and implementation processes, empower and ensure the inalienable rights of the marginalized to guide and define the values and parameters of surrounding ethical considerations. Given that many victims of discriminatory systems are often locked out from pursuing reparations, a more nontechnical formulation of technology ethics means a more generalized public understanding and control of rapidly changing machine capabilities without using ethnography as a Band-Aid after the process has been deployed. However, opening these black boxes out into the 'cold light of day' also exposes the relative incompetency of some algorithms [11]. Editorial narratives regularly resort to the formulation, "We don't know why the computer made the mistake, but it's likely because the algorithm still isn't very good at such important but difficult tasks."

3.8.1 Case Studies on Ethical Issues

In this section, we explore several case studies of AI projects, many of which were noteworthy enough to attract media attention. In these cases, suppose you had to decide how to act. What ethical principles would guide your actions? Are there any laws that are relevant to the case? If so, which? Which formal or public statements might help you decide your best action? What are the strongest and weakest arguments in favor of or against the major positions? Can you explain how these arguments might generalize to defend or criticize other situations? What would you decide if you actually had to act, taking into account what has been said previously in the forum and the lecture, discussion, and reading? To maintain the reasoning environment for future discussion, we leaving the names of our interested parties out of these scenarios. However, we remind you that the cases are based on real events [36]. Indeed, in some of these cases, you will have used the software that was created, or you might know or even have met the people who were involved, as presented in Table 3.1, with their possible effect on the humanity.

3.8.2 Mitigation Strategies

It is not all doom and gloom, though. There are certainly ways to mitigate some of the ethical and moral issues. Technological solutions can be developed with respect to "smart" tools, goods, services, organizations, and entire systems, such that the systems can be smart without being morally smart. In addition, regulatory obfuscation and a further use of transparency can be utilized. Smart goods, services, organizations, and entire systems offer benefits if they respond to the informational demands of the environment. However, to respond to the expected moral expectations of the public—without additional human involvement—these entities would need to have moral capacities. They would need to use the right judgment and make moral decisions. But based on the analyses carried out, it can be concluded that this moral expectation is hard to meet: smart systems would need to have moral mental states, and these

Table 3.1 Navigating the moral minefield using ethical dilemmas in artificial intelligence

ethical issue	AI context	Real-world example	Moral dilemma	Affected stakeholders	Suggested resolution
Algorithmic bias	Hiring systems	Biased resume screening tools	Discrimination based on race/ gender	Job seekers, companies	Diverse training data, bias audits
Privacy invasion	Smart devices	Voice assistants collecting data	Constant surveillance, consent issues	Users, tech companies	Clear consent policies, data minimization
Lack of transparency	Decision-making AI	Credit scoring algorithms	Black-box decisions with no explanations	Consumers, financial institutions	Explainable AI, open models
Deepfakes and misinformation	Media generation	Fake political videos	Manipulation of public opinion	Voters, governments	Detection tools, legal frameworks
Autonomous weapons	Military AI	AI drones for combat	Loss of human control in lethal decisions	Civilians, military	International regulation, human-in-loop
Data ownership	Predictive analytics	Health data used by third parties	Unauthorized use of sensitive personal data	Patients, researchers	Ethical data sharing agreements
AI in policing	Predictive policing	Targeting minority communities	Racial profiling and unjust surveillance	Communities, law enforcement	Transparency, community engagement
AI in healthcare	Diagnosis systems	Biased treatment suggestions	Health inequalities reinforced	Patients, doctors	Inclusive datasets, medical oversight
Job displacement	Automation	AI replacing human workers	Economic inequality and unemployment	Workers, employers	Reskilling programs, phased automation
Emotional manipulation	Sentiment analysis tools	AI-driven ads and content targeting	Exploiting human emotions for profit	Consumers, marketers	Ethical marketing standards, user control

are just not available to them. If we still want to use smart entities without added moral capacities, this means that we are willing to accept the fact that, in difficult cases, smart goods, services, organizations, and systems can disappoint us as citizens, consumers, partners, employees, or regulators [37]. In exchange for benefits, we accept that we shall not always be treated with dignity.

Although the developed ideas have covered one moral failure in the case of smart entities, namely that the entities do the wrong thing—because smart goods and services could not possibly perform their duties on the basis of a meaningless moral reflection or their environmental self-regulation broken down—there are other moral issues at stake. The focus of the chapter was on the problem of disrespect for human beings if they are being expected to perform unreasonable duties. Moreover, smart organizations and systems can have moral problems because humans do not have a clear message or role to perform anymore. They might alienate the humans who are supposed to give them input or assess their results, capacities, risk, or interest in them [38]. After all, humans want smart entities to back human performances and share the benefits of these performances, where, in the smart play of all, humans are accountable for what goes well and, although this might be a different type of duty, must be able to give an account of their actions to the smart entities that are deployed.

3.9 AI in Healthcare

Healthcare is becoming a new frontier for AI innovations, offering us the possibility to move beyond traditional population-based public health and offer prediction and intervention at the level of individuals. This means more customized and timely care. From detecting skin cancer or retinal damage to identifying white matter injury in premature babies, AI can bring the resources of the best medical centers to remote or individual patients. Diagnostic tools, especially to read radiological images, are already being deployed around the world. Expert systems can assist healthcare delivery in peripheral clinics and hospitals where access to medical diagnosis and advice is a costly proposition and requires the trained eye and judgment of a healthcare professional. Applications include chatbots, which offer basic health advice and mental health support, to AI-based help for dealing with ADHD and PTSD [9]. AI nurses can understand natural language and, as a result, can participate in interviews, collect all relevant information, and check it against the risk factor analysis system.

The use of data to improve services and outcomes is not new in healthcare. What is different is the potential for AI to do things faster, more accurately, and more uniformly and to deal with data, which comes from new sources and in different forms. Current problems with data privacy and cybersecurity need to be solved. But healthcare data have always been among the most privacy-sensitive of all. Otherwise, billions of people would not be handing the most intimate of their secrets to their medical encounters via devices that are easily lost or easily penetrated. The global market for personal health devices is potentially a significant portion of the global healthcare spend. It explores and expands in this way as governments, insurers, and health professionals search for new tools that offer flexibility, quality, and efficiency in responding to the challenges posed by an aging and changing population [15]. Debates about the obligations governments have in terms of pricing, affordability, and access to health suggest that there is also capacity for telemedicine and AI systems to offer new models of more equitable healthcare and well-being. From a critical social technology perspective or development ethics perspective, technology such as AI chatbots may engender new digital divides when it comes to healthcare. So, while these healthcare applications might serve to improve global health metrics, there is an important and currently little-studied question about whether they will also help speed up the attainment of universal healthcare [39]. Balancing both technological and institutional reform to create transformative, rather than incremental, solutions is critical.

3.9.1 Ethical Considerations

Recent advances in AI, especially in the area commonly referred to as "Deep Learning," in conjunction with vast amounts of data becoming available through the Internet infrastructure, have produced spectacular results. These include, among others, language translation, voice interfaces available on mobile devices, the ability to generate realistic images as well as deep fakes, and a broad range of intelligent behaviors. Along with these advances have come significant ethical concerns and the recognition that these technologies can be used for both good and bad purposes [40]. AI has many beneficial applications, and these can range from helping individuals with cognitive or physical disabilities to perform tasks more effectively, enhancing physical security, mitigating public hazards, or contributing to sustainable development. These potential benefits also come with potential risks and ethical concerns. Many of them are familiar from a long history of science and technology: threats to privacy, laws, and civilian rights, as well as an increasing facilitation of surveillance regimes. Given the scale of potential changes wrought by this broad class of technologies, it is clearly time to engage with this issue with a greater sense of urgency. The ethical considerations of AI and its applications do, however, introduce a collection of challenging issues [31].

3.9.2 Patient Data Security

The growing use of social robots and their sensor systems in numerous healthcare environments may have many advantages, but they also raise important concerns about patient data security. As research on the efficacy of these robots increases, potentially sensitive data about patients and medical subjects, including age, gender, hormonal state, and potentially body weight, various details about medical state, and other biometric features, may be recorded and retained. Patient data could potentially be subject to theft or abusive use if not managed in a responsible manner. Such security issues can potentially impact future patient use and adoption of social robot solutions as well, making conservative practices around patient data storage, management, and erasure crucial for social robot developers to consider [18].

In support of experimental strategies around security and data protection in healthcare settings, a patient/non-patient data distinction and research guidance labeling system on important physical and functional capabilities of robot types used with patients has been created. The importance of training or guiding future robot users in appropriate data protection practices is stressed. Different types of robots are distinguished according to their interactions with patients, their trusting relationships, as well as the sensitive data they might collect. For example, designation and use of different robot types are made clear when employed for data collection, diagnostics, or training, from robots employing tools, reminding or monitoring patients, and others that actually interact or communicate with patients as part of their therapeutic activities. For each type of patient activity, best practices for data investigation and emergency procedures guide the use of robot systems and patient protection [30]. For storage of detailed information or emergency approaches, functionality is also considered. Such differentiation allows for both mitigating potential data security fraud and for planning to share anonymized data for trustworthy robot healthcare applications under the best conditions.

3.10 The Future of AI Ethics

AI ethics will continue to be a field of innovation and debate. A critical challenge is to determine how society can ensure that all individuals, companies, and nations benefit from AI and are included in the new AI-driven world. Ethics and laws govern behavior between individuals, but many face a deeper, underlying question of what kind of society we want to collectively build. Legislation to address ethical issues will form an important cornerstone to steer the development of AI. Regulatory oversight, based on ethical principles and applied to cover the diverse industries responsible for its development, will offset the risks associated with detrimental bias and actions. Industry also will play an important role in fostering transparent, reliable, and trustworthy AI systems and helping to advance efforts. We must be visionary and take maximum advantage of this potential to address the greatest challenges and opportunities in the twenty-first century [17]. The full potential of AI will only be reached when we can ensure widespread trust in the AI systems supporting our industry, government, and society. We have an ethical imperative to research, develop, and use AI responsibly and for the benefit of one and all. Just as recent advances in AI owe the existence of its predecessors, our collective future advancements will transform the world around us, as the technology has done since the Stone Age. In modern times, advanced technology can create new job opportunities, foster greater efficiency in the workplace, and enhance experiences for individuals in their daily lives [41]. Through the ethical pursuit and dissemination of AI knowledge, we can help ensure a future where AI technologies work for us and benefit not just a few, but each and every one of us. Thank you.

3.10.1 Emerging Technologies

Emerging technologies have enormous potential for influencing the character, form, and destiny of human society. Historically, the introduction of new technologies has created significant changes in human society, but it has also contributed to disruption, cultural and physical decay, and diminished human values. Society needs to guide the development of new technologies and constrain their potential for misuse, just as emerging technologies ought to be informed by societal needs and values. These needs are particularly significant with respect to the development of technologies that have a pervasive impact on intrinsic human values, such as security, privacy, human relations, or personal self-worth in the context of work [8]. A series of policy issues arise around the "responsible development" and application of new technologies in these domains, issues that challenge our existing regulatory structures.

The rapid and accelerating pace of technological evolution in the areas of computer and communications technologies has created a unique set of problems and opportunities. The technology appears to evolve so rapidly, in fact, that any extension of or change to the existing regulatory environment may be outdated, counterproductive, or even entirely moot by the time it becomes effective; a coherent policy structure is simultaneously subject to the law of unintended consequences from a rapidly modifying technological environment. This suggests the necessity of considering guidelines for "responsible development" in those areas of research in which technology will create the most pervasive impact on human values. Given the potential impact of advancing technology on personal privacy and personal dignity in the context of the workplace, including threats to humans from de-skilling or displacement from work, and the erosion of societal trust and security, a critical group of interested parties are the stakeholders who are affected by the research [13]. These stakeholders include technology developers who can promote thoughtfully developed messages that can encourage both careful attention by technology investigators to the social consequences of their work and feedback from public interest groups. All formal organizations engaged in technology research should develop and express ethical guidelines declaring a commitment to avoid behaviors that will erode societal values. These organizations should receive appropriate public and private benefits in recognition of their adherence to the declared ethical standards [39].

3.10.2 Global Standards

We need an international conversation where the big players get together. The good news is that tech companies have also started taking this idea seriously. The organization is keen to explore issues arising from the filter bubble, potential unemployment caused by technological developments, and bias in AI algorithms. These are good signs that the tech giants are actually paying attention. Still, there is a significant need for international bodies to take on these challenges because these organizations have the most experience, credibility, and reach. It is important that this is coupled with transparency on the part of tech companies in order that the broader community can have access to more of what these companies know [15]. The second part of the equation is that smaller and medium companies are here to stay and are also eager to get involved. This is good news: globalization brings responsibilities.

Globalization means we have to play by a set of rules that grows wider. The international playing field is still too often characterized by rogue entities that are happy to ignore some sorts of rules if they bring an unfair competitive advantage, and that's always going to be true. There are methods for encouraging standards in the hope that competition will reflect the highest common values of conduct. These are voluntary commitments. There is nothing that requires a national body to adhere to a set of global standards, but there's a big benefit to aligning their industry with what consumers are asking for. Technology companies tap into humanity's desires and dreams, so they have an incentive to make the ideas of others more concrete. The tech companies all want to be major players on the international scene, and they understand that this objective needs trustworthy, accountable AI, along with societies that understand how powerful AI can be [14]. The challenge now is to continue to create the momentum for durable global structures and commitments that encourage the use of AI for the good of humanity.

3.11 Public Perception of AI

The ethical concerns for AI and its public perception are intertwined. As AI makes more decisions by itself in social settings, public perception becomes increasingly important. From the perspective of the three ethical dimensions that I mentioned earlier, public perception seems to be a minor problem in distributive justice. AI products are just objects bought by consumers like any other products. The measure of how willing consumers are to purchase these products seems to reflect public opinion on distributive justice. However, AI's impact is more significant than most other products. All individuals in society are stakeholders, even if they never purchase or use AI products [11]. The public perception of social justice and species justice is the most essential foundation for these two kinds of ethical justice. Public perception of social justice, from the fairness of procedures, has been essential for AI creators to consider. The underlying premise is that if we can deal with public concern satisfactorily, the implementation of AI is implicit with a demand for societal benefit. Therefore, beneficence is a consensus of future work. It is an open question whether the threshold of societal benefit is a moral prerequisite or if policies dictate it. With the proper implementation of AI fairness tools using procedures and algorithms favoring fairness objectives, the thoughts of many who now do not trust AI, including those fearful that it will displace a large fraction of the workforce, will slowly evolve [18]. Since the launch of the conference on fairness, accountability, and transparency in 2017, AI ethics has received much more attention than before. The improvements of many algorithms are based on the relaxation of fairness theories that have been used in AI rapidly.

3.11.1 Misinformation

Perhaps the most pervasive ethical dilemma associated with AI is the issue of misinformation. AI itself is not the root cause of misinformation, but rather accelerates the process of curation. The speed and ease with which inaccuracies are propagated online has led to a phenomenon uniquely associated with this technology, often referred to as deepfakes. Deepfakes refer to AI-generated synthetic media algorithms used to falsify audio and video content. Developing from a combination of the phrase deep learning and the word fake, deepfakes have become notorious due to the very real threat to privacy and security that they pose. From malicious political propaganda to actors in the adult industry, deepfakes can compromise every aspect of society. The fact that deepfakes can take the face of an individual, superimpose it on someone else, and then mimic their facial expressions has raised significant concerns. We have become more open to believing the veracity of information we receive digitally. Likewise, we are becoming less equipped to confidently identify the signs of a deepfake [35]. With calls for AI giants to systematically detect deepfakes increasing, the issue has reached corporate and governmental attention. However, creating this safety net and defeating the sophistication of deepfake technologies remains an immense challenge.

3.11.2 Education and Awareness

One of the biggest issues to address in the transition to a world where AI plays a much larger role is that of public understanding of the technology. AI is not understandable to the public in the same way that cloud computing might be. This is not necessarily the public's failure. As we have mentioned, AI is not one big groundbreaking technology; rather, it is a multitude of related technologies that solve problems generally thought to require intelligence. When these technologies improve, we say that AI is getting better. This leads to the second problem: public discourse on AI often defaults to could scenarios—we all default to worst-case scenarios—rather than would or more likely scenarios. Stray further into advertising and even the would becomes can [31]. The challenge is to not oversimplify our understanding of AI but to ever strive to delve into it. A simple headline about a lost kitten in an AI-powered washing machine scares not only that the writer takes the worst possible scenario as a given (an assumption that requires a high burden of proof from a technical perspective), but also that washing machines will suddenly have AI in them. The panic that an AI-induced emotional crisis inflicts on a criminal career is based on a misunderstanding of categorization as AI-driven software agents. This lack of understanding creates a dangerous imbalance of trust—it is rooted in a misunderstanding of cause, capability, and intent [3]. AI researchers, companies, and governments need to work on AI education and awareness if we are to plunge headfirst into the oncoming tide of intelligence.

3.12 Case Studies in AI Ethics

In this section, we consider two specific case studies in the deployment of AI systems at scale: the use of risk assessment algorithms in the criminal justice system and the use of algorithms in content moderation on social media platforms. Perhaps at the forefront of the public consciousness of AI ethics are questions surrounding its use in maintaining or enforcing the authority of the law. A particularly prominent instance has been the application of risk assessment algorithms in the context of the criminal justice system. A private company offering risk assessment tools making recommendations for judges, probation officers, and case managers produces risk assessment scores in criminal sentencing and re-entry. The use of these scores by judges in criminal cases has become the subject of much public scrutiny [37]. These tools use longitudinal data over trial defendants, interpreted through machine learning techniques, which correlate outcomes such as recidivism with the behavior of earlier defendants and could predict the risk that a defendant currently before the court would reoffend in the future.

3.12.1 Real-World Examples

There are countless examples of potentially harmful uses of AI that lie waiting in various fields. From employment to defense to entertainment to medicine, the uses of AI are so varied that trying to predict all the ways in which AI could go wrong is itself a daunting task. While all ethical discussions of AIs are largely speculative at this stage, and our purpose here is to provide a framework for considering these ethical issues, there is a great deal to unravel in considering a variety of potential ethical dilemmas. These are real-world examples the confluence of AI and ethics is faced with, often in the immediate future. Ministers of defense and country leaders have threatened and taken the AI arms race to the next level. The navy have declared that they fully expect to have AI-controlled 3-D printed drones within two years, and the military is exploring converting surveillance drones for combat purposes [36]. The fact that a state-sponsored actor recently shot down an American surveillance drone serves as a strong reminder that the relatively benign and ethical use of AI for safety purposes is unlikely to be coupled to a worldwide ban on disagreeable uses of AI. Remote nuclear silos continue to be a flashpoint and a possible mistranslation between intent and understanding may no longer be an unlikely far future prospect, but instead be a significant problem for humanity that needs to be addressed. Even advancements in cloud infrastructure can be weaponized, as an organization revealed the possible misuse of cloud computation for mass surveillance [42].

3.12.2 *Lessons Learned*

We hope the informal treatment to four of the key ethical issues may stimulate thinking and discussion on the identification, avoidance, and resolution of moral issues in the design and use of AI systems. These issues have no easy answers and general advice such as "be careful" and "don't be complacent" does not help much when hard decisions are to be made. It is necessary to recognize that failures can occur and to design systems and organizations in a way such that incidents are both less likely and better handled when they occur. This is a psychological challenge: the formal approach avoids ethical considerations but reduces the psychological tension. Examples of failures from many sectors, not just AI, can be used to raise awareness and stimulate discussion on how problems may be overcome [43]. The incident reporting scheme in medicine has not solved all the problems in that domain, but it has had a considerable impact on raising the perceived importance of errors, and will continue to do so. The AI and law communities have demonstrated quite close involvement in using specific major failures as the basis for detailed design of and requirements specification for AI and computer law systems and for the development of an adversarial legal process that can deal with policies and corrected laws when necessary. By fostering a certain minimum diversity and open-mindedness in the users of AI, the legal system helps curb potential harm that could result from over-reliance on fully automated, rule-based AI [6]. In other words, the creation of a legal environment and the involved use of complex systems provide some of the recursive control mechanisms a complex system like human society needs to contain the risk of catastrophic consequences resulting from the incomplete understanding or badly tuned AI-based systems.

3.13 Interdisciplinary Approaches to AI Ethics

AI technologies are diverse, and their ethical, legal, and social implications are complex. Interdisciplinary AI ethics is inherently interdisciplinary, with contributions arising from computer science, data science, software engineering, social sciences, ethics, philosophy, psychology, neuroscience, law, and policy studies. Ethically aligned AI systems must be flexible and thus able to accommodate contributions making use of different theoretical approaches from different scientific fields. If we wish to build ethically aligned AI, we must engage in diverse interdisciplinary research and innovation. AI is a science and a technology, but AI research and AI technologies are also profoundly shaped by ethical, legal, and social influences [13]. Creating ethically aligned AI systems is challenging for the following reasons: AI makes high-stakes decisions; AI reflects and reinforces biases; AI shapes human behavior and social institutions; AI has uncertain effects on global stability. The study of artificial morality is not itself a new field. It is notable, however, that contemporary theories of artificial morality and accounts of interdisciplinary AI ethics often draw upon diverse fields including computer science, data science, software engineering, social science, epistemology, ethics, philosophy, psychology, neuroscience, law, and policy studies. Accordingly, in this chapter we use the term "interdisciplinary" liberally to encompass joint research across these contributing fields [9].

3.13.1 Collaboration Between Fields

Thus far, we have looked mainly at the task assigned to artificial intelligence systems—the landscape of human decision making that these "robots" are wandering into. However, a collaborative framework often extends closer to home. AI is not just a contributor and a recipient; it can often become the mechanism that enables work between different stakeholders. Machine learning techniques and large data sets often allow us to zoom out a level, focusing not just on specific strings of language or decisions, but on the meta-problems: the hidden incentive or systemic failures that underlie a problem to begin with. This allows us to abstract away from technical and institutional silos—to consider seemingly disconnected phenomena in the same analytic framework—let alone solve them [32]. Fields across the social and physical sciences have invested significant intellectual and technological effort in using computational methods, particularly those that are associated with machine learning, to do more with their data. Collaborative initiatives have been launched in numerous traditional fields such as the social sciences, economics, computational biology, developmental economics, general biology, political science, history, geography, and public health. Moreover, our goal is to encourage these different fields to engage with AI in a way that goes beyond isolated one-off uses of AI methods [10]. These fields join a longer history starting in the late 1970s, in essence with the creation of the field known as "artificial intelligence."

3.13.2 Role of Social Sciences

Ethics is not the only field that is exposed to complex engagement with AI. Indeed, many of the classic concepts of academic life could be transformed. Economists will no longer be able to assume a common-sense foundation generically available to intelligent agents, leading them to different techniques of analysis. Psychologists will be able to design virtual environments to explore the limits of human cognition. Political scientists can explore interesting institutions and the limiting hypotheses of different political systems. Lawyers can ask questions about the legal status of agents that inhabit distributed systems and of those systems themselves. And this list is by no means exhaustive [12]. Physicists engaged in quantum computing should take a lesson from deterministic chaos, that they will never be able to construct a perfect model; it will always be corrupted by every stray hamburger in every location between the research library and the research lab. Those who would expect predictability should abandon such models.

Social scientists have a serious task before them in understanding the evolution of agents in interactive systems. They need to understand the nature of the distributed software hierarchies that are being created and their likely interactions. The research reports on an anthropological vision of the market used as a simulation for exploring the emergence of institutions under different rules of engagement and opens up the conclusions to a wider range of systems. Such research is of marginal relevance to the urgent short-term problems posed by the development of fixed, stand-alone AI systems and even distributed collections of such systems [14]. What is important is to recognize why ethical constraints are necessary, to operate at an appropriate distance from the problem, and to start exposing such developments to a wider audience. The research proposes two broad ethical considerations: that the minimum overriding consideration is the humane treatment of agents as intelligent systems; that proliferation must be limited and the application of forbidding weapons to mayhem discouraged. We applied statistical theory to provide bounds on the likely development of market institutions whose purpose was to maintain order inside industries where history, skills, or both could not easily be marketed at a price that would pay for training new generations [11, 39].

3.14 Conclusion

The ethical issues raised by AI are profound and far-reaching. In this field, as in life, there are no ethical or unethical technologies. Instead, what we have are human artifacts and human decisions about the uses to which they are put. AI researchers and other stakeholders, such as philosophers, policymakers, and industry leaders, should reflect closely on the ethical implications of their work. AI researchers can address the question of what we ought to do in their work by consulting theoretical moral philosophy, constructing moral theories that can be implemented through codes of ethics and software operationalizing the ethical behavior those codes mandate. Important domain-specific moral questions about the AI used in, for example, healthcare or weapons deployment, are elicited from those who practice those professions. In this chapter, we have sought to set out the ethical implications of the use of AI. Although it may not be its primary function, the potential harm that AI can cause is significant, and it is important to try to mitigate that risk as much as possible. In order to ensure that AI is aligned with ethical principles, its practitioners, policymakers, and society need to understand what those principles are and how to fulfill them. The involvement of AI in our lives presents us with significant new questions about humanity. These ethical dilemmas are challenging us to understand anew what we mean when we talk about good or evil, justice, fairness, right, wrong, virtue, and the expression of freedom in a modern, networked, high-technology society. At every stage, AI researchers must disclose the uncertainties and stakes involved, allowing people to make their own personal, moral, and social judgments not just in the laboratory but as democratic and civic actors, and then to make AI do and think as the people want.

References

1. Anshari M, Almunawar MN, Masri M, Hrdy M. Financial technology with AI-enabled and ethical challenges. Society. 2021;58(3).
2. Shafik W, Kalinaki K, Fahim KE, Adam M. Safeguarding data privacy and security in federated learning systems. In: Federated deep learning for healthcare [Internet]. Boca Raton: CRC Press;2024. p. 170–90. https://doi.org/10.1201/9781032694870-13.
3. Rezwana J, Maher ML. User perspectives on ethical challenges in human-AI co-creativity: a design fiction study. In: ACM International conference proceeding series. 2023.
4. Christou P. How to use artificial intelligence (AI) as a resource, ow methodological and analysis tool in qualitative research? Qualitative Report. 2023;28(7).
5. Shafik W. Deep learning impacts in the field of artificial intelligence. In: Deep learning concepts in operations research [Internet]. New York: Auerbach Publications;2024. p. 9–26. https://doi.org/10.1201/9781003433309-2
6. Shafik W. Ethical and legal considerations in artificial intelligence. In: AI-Enabled threat intelligence and cyber risk assessment. 2025;90.
7. Johnson J. The AI commander problem: ethical, political, and psychological dilemmas of human-machine interactions in AI-enabled warfare. J Mil Ethics. 2022;21(3–4).
8. Casas-Roma J, Conesa J, Caballé S. Education, ethical dilemmas and ai: from ethical design to artificial morality. In: Lecture notes in computer science (including subseries lecture notes in artificial intelligence and lecture notes in bioinformatics). 2021.
9. Peckham JB. The ethical implications of 4IR. J Ethics Entrep Technol. 2021;1(1).
10. Uddin ASMA. The era of AI: upholding ethical leadership. Open J LeadShip. 2023;12(04).
11. Nassar A, Kamal M. Ethical dilemmas in AI-powered decision-making: a deep dive into big data-driven ethical considerations. Int J Responsible Artif Intell. 2021;11(8).
12. KOMPELLA K. The trolley problem and ethical dilemmas in AI. Inf Today. 2020;37(5).
13. De Gagne JC, Hwang H, Jung D. Cyberethics in nursing education: ethical implications of artificial intelligence. Nurs Ethics. 2023.
14. Gouvea JS. Ethical dilemmas in current uses of AI in science education. CBE Life Sci Educ. 2024;23(1).
15. Vousinas GL, Simitsi I, Livieri G, Gkouva GC, Efthymiou IP. Mapping the road of the ethical dilemmas behind artificial intelligence. J Polit Ethics New Technol AI. 2022;1(1).
16. Shafik W. Generative artificial intelligence for social good and sustainable development. In: Generative AI: disruptive technologies for innovative applications. 2025;153–85.
17. Song X. A study of ethical dilemmas and regulation of AI chatbots. J Theory Pract Soc Sci. 2023;3(9).
18. Guan H, Dong L, Zhao A. Ethical risk factors and mechanisms in artificial intelligence decision making. Behav Sci. 2022;12(9).
19. Nassar A, Kamal M. Ethical dilemmas in AI-powered decision-making: a deep dive into big data-driven ethical considerations. Int J Responsible Artif Intell. 2021
20. Shafik W, Kalinaki K. Societal and ethical implications of technology-enhanced agriculture and healthcare: an African context. In: 2024 IST-Africa conference (IST-Africa) [Internet]. IEEE;2024. p. 1–11. https://ieeexplore.ieee.org/document/10569306/

21. Shafik W. Generative adversarial networks: security, privacy, and ethical considerations. In: Generative artificial intelligence (AI) approaches for industrial applications. Springer;2025. p. 93–117.
22. Shafik W. Toward a more ethical future of artificial intelligence and data science. In: The ethical frontier of ai and data analysis [Internet]. IGI Global;2024. p. 362–88. https://doi.org/10.4018/979-8-3693-2964-1.ch022.
23. Shafik W. Quantum computing and generative adversarial networks (gans): ethics, privacy, and security. In: Quantum AI and its applications in blockchain technology. IGI Global Scientific Publishing;2025. p. 111–56.
24. Shafik W. Generative AI for social good and sustainable development. In: Generative AI: current trends and applications. Springer;2024. p. 185–217.
25. Abubakari MS. The internet of things (IoT) as an emerging technological solution for the covid-19 pandemic mitigation: an overview. In: Khairudin M, Asnawi R, Djatmiko IW, Sudira P, Hadi S, Arifin F, editors. J Phys: Conf Ser [Internet]. IOP Publishing Ltd;2021. https://www.scopus.com/inward/record.uri?eid=2-s2.0-85101857842&doi=10.1088%2f1742-6596%2f1737%2f1%2f012003&partnerID=40&md5=09b21844dcc4f91d84d2e67880505d83
26. Paraman P, Anamalah S. Ethical artificial intelligence framework for a good AI society: principles, opportunities and perils. AI Soc. 2023;38(2).
27. Shafik W, Zakari RY, Kalinaki K. Ethical and privacy concerns in bioinformatics and cyber-physical systems integration in healthcare. In: AI-Driven personalized healthcare solutions. IGI Global Scientific Publishing; 2025. p. 333–64.
28. Miao J, Thongprayoon C, Suppadungsuk S, Garcia Valencia OA, Qureshi F, Cheungpasitporn W. Ethical dilemmas in using AI for academic writing and an example framework for peer review in nephrology academia: a narrative review. Clin Pract. 2024;14
29. Shafik W, Singh R, Kumar V. Artificial intelligence transparency and explainability in sustainable healthcare. In: Transforming healthcare sector through artificial intelligence and environmental sustainability. Springer;2025. p. 165–91.
30. Oniani D, Hilsman J, Peng Y, Poropatich RK, Pamplin JC, Legault GL, et al. Adopting and expanding ethical principles for generative artificial intelligence from military to healthcare. NPJ Digit Med. 2023;6(1).
31. Kooli C. Chatbots in education and research: a critical examination of ethical implications and solutions. Sustainability (Switzerland). 2023;15(7).
32. Baum SD. Social choice ethics in artificial intelligence. AI Soc. 2020;35(1).
33. DataEthics. Addressing ethical dilemmas in AI: listening to engineers. Association of Nordic Engineers. 2021.
34. Arroyo P, Schöttle A, Christensen R. The ethical and social dilemma of AI uses in the construction industry. In: IGLC 2021—29th Annual conference of the international group for lean construction—lean construction in crisis times: responding to the post-pandemic AEC industry challenges. 2021.
35. Jabotinsky HY, Sarel R. Co-authoring with an AI? Ethical dilemmas and artificial intelligence. SSRN Electron J. 2022.
36. Coltri MA. The ethical dilemma with open AI ChatGPT: is it right or wrong to prohibit it? Athens J Law. 2024;10(1).
37. Strümke I, Slavkovik M, Madai VI. The social dilemma in artificial intelligence development and why we have to solve it. AI Ethics. 2022;2(4).
38. Zhang Z, Chen Z, Xu L. Artificial intelligence and moral dilemmas: Perception of ethical decision-making in AI. J Exp Soc Psychol. 2022;101.
39. Machado H, Silva S, Neiva L. Publics' views on ethical challenges of artificial intelligence: a scoping review. AI Ethics. 2023.
40. Rahayu I, Ardiyanti H, Judijanto L, Hamid A, Bani-Domi ES. Ethical dilemmas and moral frameworks: navigating the integration of artificial intelligence in Islamic societies. Int J Teach Learn (INJOTEL). 2023;1(3).
41. Shafik W. Artificial intelligence and machine learning with cyber ethics for the future world. In: Future communication systems using artificial intelligence, internet of things and data

science [Internet]. Boca Raton: CRC Press;2024. p. 110–30. https://doi.org/10.1201/9781032648309-9.

42. El-Deeb A. Behind OpenAI CEO dismissal: an ethical dilemma and a new AI revolution. ACM SIGSOFT Softw Eng Notes. 2023;49(1).
43. Prikshat V, Patel P, Varma A, Ishizaka A. A multi-stakeholder ethical framework for AI-augmented HRM. Int J Manpow. 2022;43(1).

Chapter 4
The Future of Work: AI's Threat to Jobs and Economic Stability

4.1 Introduction

Artificial intelligence is an extension of automation, which has been around for centuries. Mechanization involved replacing human-operated tools with machines. Automation went a step further, using feedback mechanisms to create a machine that could perform a series of tasks by itself [1]. A modern example is a pilot program that successfully flew surplus fighter planes converted into drones. Automation uses programmed responses because the tasks it is built to perform are routine, meaning they are predictable and can be performed following a series of commands [2]. Conventional automation can replicate only well-defined tasks. AI-based automation mimics human thought processes and learns by doing, recognizing and abstracting information so it can deal with a new, non-routine situation. Today, AI and automation have a broad range of capabilities. Development is becoming ever more sophisticated. Types of AI and automation include robotics, machine learning, deep learning, physical AI, cognitive AI, decision management to include functions such as problem identification and decision-making, and an AI platform workbench that makes applications possible [3].

4.1.1 *Understanding Artificial Intelligence*

Artificial intelligence can be defined as a set of technologies that enable a machine-driven system to imitate and extend human cognition. While a narrow or "weak" version of artificial intelligence may perform a very specific task, a broader or "general" version has the capacity to perform a wide range of tasks that require human-like cognitive abilities. "Learning" is often added to the definition because, in practice, the performance of these systems becomes more predictable and variable over time in the language in which they are trained. This can be constructed with or without

W. Shafik, *The Dark Side of AI*, https://doi.org/10.1007/978-3-032-09130-7_4

human supervision and ways to achieve human tasks with increasing efficiency and effectiveness. AI can be defined as a constellation of technologies informed by design philosophies that emphasize learning, adaptation, and goal-directed behavior. Taken together, these characteristics can create systems that perform, generalize, and act in ways that would appear to be tasks performed by humans, often with greater ease and at lower economic costs [4]. Moreover, AI has a broader variety of applications than previous computing techniques. Its potential spread of economic activity is substantial.

4.1.2 Types of Automation

Automated machines have been replacing human labor for centuries. Most jobs previously performed by humans have a potential for automation. However, the timeline for this development has dramatically shortened with AI. AI systems can now perform tasks in complex problem-solving, such as strategy games, even better than humans. The question is not so much whether AIs can perform tasks as competently as humans, but which tasks and at what cost? Oddly enough, despite a variety of ways to automate jobs, certain jobs have actually increased in their complexity and demand for human involvement. As mechanical tasks are automated, more specialized human services have developed to accomplish the tasks that complement automation [5]. In this chapter, we adopt the following taxonomy for different types of automation: routine versus non-routine, manual versus cognitive. Routine tasks are those that are well-structured and repetitively performed, where sequences of actions can be written down and programmed. Cognitive tasks are generally more difficult to automate due to their high level of variability, requiring both knowledge and situational understanding, such as common sense and emotional intelligence. Manual tasks, being more repetitive or mechanically repetitive, are easier to automate compared to cognitive tasks. Prior to the arrival of AI, manual tasks occupied the forefront for doing work that was more suited for mechanical automation [6]. The recent advances in AI have broadened automation possibilities to cognitive tasks, which are commonly believed to be reserved for human workers.

4.2 Historical Context of Work and Technology

The replacement of workers by machines has been a critical part of economic and social evolution since the first industrial revolution. In 1564, for example, Queen Elizabeth I denied a patent to William Leemings, who had invented a stocking machine that threatened to disrupt the established community of hand-knitters in London. When Charles I was unwise enough to skimp on the payment of his taxes to support inflationary wars in Scotland, he renewed a uniform disability to all coming from elsewhere and imposed a tax on anyone wearing silk or satin [7]. The recipients of these concessions were the machines and their masters, whose nomination as official manufacturers of men's clothes and ladies' shoes enforced the ruin of activities that had once been the pride of London's craft population. The process of "creative destruction" is an essential element of capitalist dynamism and has been a central theme in the evolution of economic history. The expansion of commerce in the mercantile period, the dispersion of factories after the industrial revolution in Great Britain, and the large outcry against the separation of "master and men" in the decades that followed were three well-documented examples of this process [8]. These are also the most striking examples, but certainly not isolated.

There is nothing new, therefore, in the introduction of new machines that perform tasks formerly carried out by human workers. There is nothing new in the criticism of technological change, as if it were always an avoidable moral misstep rather than an evolution in the structure of production that, despite all the difficulties, has always been the main source of wealth creation and progress [9]. What is new today about this phenomenon is that an expansion of demand follows the classic story of machines providing an increase in production, and an enrichment of the previous workers no longer seems to apply. A second, more serious consideration is that very advanced technologies reach spaces that, until now, man had considered safe from their penetration. This observation revolves around the concept of Kleiber's law relating the metabolism of a living being to its mass [10]. The law of metabolism indicates that the number of people that the fruits of the earth can feed is not proportionate to the square, but to the population of the global population.

4.2.1 The Industrial Revolution

The first industrial revolution harnessed steam-driven engines to replace both human and animal muscle power with machines, leading to a near doubling of per capita income in European countries between 1800 and 1871, and then to the dissemination of this technology leading to a corresponding improvement elsewhere [11]. In more recent decades, these forces of technological advancement have been turned on the service sector: for example, algorithms do a lot of the trading of financial assets. But progress proved slower, because machines can easily displace human muscle power, but their ability to replace much of human brain power remains limited, and history as it has played out mostly involves humans creating complementary tools for machines, not being supplanted by them [12]. Perhaps the most famous 'challenge' to the nature of work came in 1931 when it was noted that 'We are being afflicted with a new disease of which some readers may not yet have heard the name, but of which they will hear a great deal in the years to come—namely, technological unemployment. This means unemployment due to our discovery of means of economizing the use of labor outrunning the pace at which we can find new uses for labor.' Since these lines were written, and despite massive improvements in labor-saving technologies, the pace at which automation has displaced human labor has been matched by productivity-driven economic growth. New technologies have led to the end of jobs and activities that were commonplace not so very long ago [13]. However, at the same time, these developments have led to the creation of new activities and jobs. In the language of classical economics, the positive impact of automation on growth has been productive: it has led to improvements in the human condition.

4.2.2 Technological Advancements in the 20th Century

The ability of artificial intelligence to perform complicated tasks has led to panic surrounding AI's job displacement potential. But it is actually the rise of the technology behind AI long before the intelligent facets entered the picture that has drastically threatened jobs in certain sectors. Many believe that the second half of the digitization of knowledge, including text, audio, and image, is what is truly threatening jobs across all skill levels. Stand-alone industries have been born from the result of these conquests, even shaping businesses and industries not traditionally considered digital. Many technologies today couldn't have been created without the invention of electricity and our understanding of materials [14]. In today's age, large breakthroughs in the digitization of knowledge have created several tasks for machines that were previously only able to be handled by humans. The creation of machine learning relations has given machines the ability to take on non-routine tasks, and if they exhibit the same level of performance or even close to it, they will be passed over for human-holding tasks. We have recently only fought the difficulties of social science, but the rise of machine learning and AI has led to a new wave of excitement when thinking about what robots, AI, and automation are capable of as the future of work comes to light, paralleling what the advancement of electrotechnology has given us, including microelectronics and computing in the twentieth century [15].

4.3 Current Trends in AI and Employment

Current trends in AI and employment indicate that we are unlikely to see the mass unemployment predicted by both classical and neoclassical economists. This is due to the fact that since 1995 and the diffusion of the internet until 2003 and accelerating until 2008, there was no increased technological displacement of employment. Also, during the last general recession (2008–2016) we note much increased technological displacement either. Most of the replacing occurs in manufacturing, transportation, communication, and finance. We still have potential replacement of many jobs into the future. Indeed, a study projects 47% of all US jobs are at risk in the next 20 years. A study had an even higher percentage of about 55% of jobs lost to automation; however, in their paper, they discuss that about 9 million jobs would be created because of the automation of driving, and socially, those jobs will probably create more than about 9 million [16]. Still, most of the jobs at risk are low-wage paying jobs that people depend on for their income, medical insurance, and retirement.

4.3.1 *Job Displacement Statistics*

The US industrial base lost 5.7 million manufacturing jobs between 2000 and 2010 and has lost jobs during nearly every year since 1979. At the same time, other countries began to enjoy sharp gains. Between 2000 and 2004, the US share of computer and electronic products manufactured worldwide dropped from 31 to 25 percent. However, US productivity reached new highs, and the country's dependency on manufacturing jobs decreased. The total private service industry added 15 million jobs from 2001 to 2010, though in developing countries, the number remains low [17]. Some economists began to suggest that although initial job losses caused by the movement of manufacturing overseas caused some harm to specific areas of the US, aggregate welfare gains from economic free trade would more than make up for the localized downsides through consumer savings and other gains. While the conversion from manufacturing workers to service industry workers is a possible positive change, job restrictions may slow this conversion and limit economic welfare gains [2]. The conversion from manufacturing jobs to service sector jobs is clearly seen in our economy. In 1990, a larger number of people worked in manufacturing and agriculture than in leisure and hospitality, professional and business services, trade, transportation and utilities, government, education, and healthcare together. In 2010, more people were working in the six service industries mentioned above than in any of the manufacturing and agriculture industries. The loss of these manufacturing jobs often created branch plants where manufacturing had moved. Such was the case in Greeneville in the state of Tennessee, which was one of the locations to which manufacturing of television tubes moved, allowing for the manufacture of glass television tubes prior to the job losses. Since the Greensville plant only paid $5 per hour compared to the $26 per hour wage paid at the New Jersey Toshiba TV glass plant that closed, TV companies were able to advertise the cost savings in a seemingly painless exchange for TVs for larger manufacturing losses [4].

4.3.2 *Industries Most Affected by AI*

One of the primary industries affected by automation will be the transport and storage sector due to the popularity of driverless cars. Moving to a self-driving economy also means transferring the human workforce to other professions. Many people who currently drive for a living would struggle to compete in a labor market with a surfeit of retail assistants and food service jobs. The electronic assistant's industry will impact many sectors like retail, administration, and service, as the work traditionally assigned to humans will gradually transition to computers. There will,

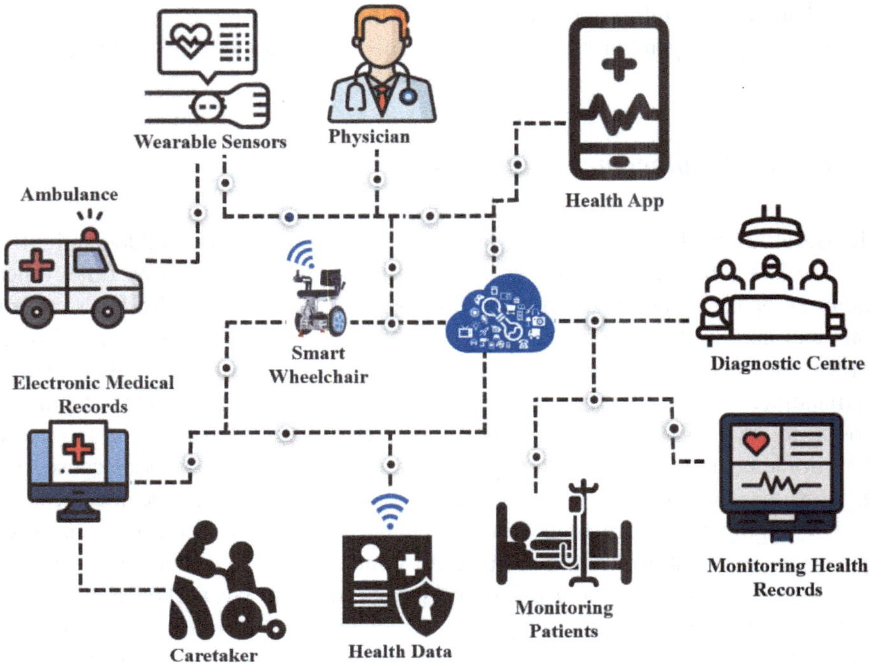

Fig. 4.1 Application scenario of artificial intelligence in healthcare

however, be new non-electronic professions in which humans will continue to outperform machines in areas such as political lobbying, dispute settlement, or marketing. Another industry to be affected is the insurance business [18]. Big data combines with predictive analysis through cloud computing, and this encompasses activities that deal with repetitive functions and rules. Similarly, transaction processing in domains involving credit checks, disputed insurance claims, automated bargaining, retailing, financial products and services of all types, or risk assessment is moving from labor to computers [5]. In particular, the impact on Western economies of the offshoring of business processes and service provision to other regions, with the loss of jobs in the West, can also be quite significant, contributing to socio-economic instability beyond its application in healthcare, as illustrated in Fig. 4.1.

4.4 Economic Implications of AI on Labor Markets

One difficulty in understanding the future of work stems from the wide range of expectations generated in the common current economic terms. Labor economists tend to examine data that reflect the past. This is useful because the historical data can at least be accompanied by an accounting of the errors in past projections. The consensus of previous studies is that the U.S. labor market has, since 1970, been primarily oriented around high-skill jobs. However, many forecasters now predict that AI will displace many high-skill jobs. This might well be true, but the forecast requires a brief description of the structure of the U.S. economy since 1970 to be essentially very different from what had existed previously (and largely what still exists in the U.S. today) [6]. Policymakers have always responded to labor market trends and fluctuations. They have promoted investment in infrastructure, facilitated capital formation, and developed business lending. But the present debate largely revolves around safety nets. Public policy is currently designed to catch people after the labor market fails them. There is a universal worry across political and ideological lines that the existing safety nets cannot or will not prevent a meaningful fraction of those affected by the transition to AI from falling into a deep crevasse of financial instability. The debates around this concern remind one of the debates surrounding the invention of the county poorhouse in the mid-nineteenth century. There seems to be collective uncertainty as to what sort of collective response there should be [8]. Yet no single entity can be expected to pay for the deep government involvement required.

4.4.1 Impact on Wages

AI, like any technological advance, will affect the incomes of different segments of the labor market differently. AI is likely to boost economic performance and stimulate the creation of new businesses and industries, which will create new jobs. With output rising, there is an implicit long-run rise in the wage bill, which means wages will, on average, rise. That said, this is an average measure, and with these additional wages highly concentrated away from the areas that the AI revolution benefits, many workers may experience falling real wages. In the long run, all citizens benefit from lower prices and better goods. Over time, workers who adjust successfully will benefit from the creation of valuable new insights, roles, and fields of work. Indeed, the natural advantage for human workers is their intelligence, which opens the door to many roles that cannot be automated [19]. And given that productivity increases as technologies complement rather than substitute for people, these roles can contribute uniquely to productivity gains in fast-adopting organizations. That, in turn, suggests that the most competitive sectors and firms will be able to afford more labor, thereby creating increased demand for workers across the economy. However, the benefits of AI to firms, innovation, and society at large are unlikely to happen automatically and evenly. Initiatives from both the public and private sectors are necessary to ensure that the transition to AI enhances the welfare of individual workers [20].

4.4.2 Changes in Job Quality

Job losers in middle-skilled positions who find re-employment more often take jobs in occupations in which they earn less than before job displacement. These findings echo an earlier study using a different data set to compare the wages of displaced manufacturing workers in new jobs with their wages prior to job displacement. They found that earnings losses were pervasive, both among those who found re-employment and among those who experienced a spell of unemployment prior to re-employment [21]. The earnings losses were especially large for workers who had held high-wage jobs prior to job displacement. For the older worker group in their examination, between the time workers took their newly held post-displacement jobs and the mid-1990s, the earnings of these re-employed workers showed only marginally significant annual increases compared to their earnings at the time they were displaced [22].

The wage losses for some displaced workers are particularly sobering when viewed from the standpoint of the share of American workers ages 25–34 and 40–49 without a 4-year college degree employed in high-, middle-, and low-skilled occupations. The value for the age group and skill category combination reported on each line represents the annual average over the entire period studied. Researchers have categorized workers into high-skilled, middle-skilled, and low-skilled occupations based on the mean earnings of workers in those occupations. On an average annual basis from 1980 to 1988, between 25 and 31 percent of the workers employed in a high-skilled occupation in the higher skill group lost their jobs at some time during the subsequent year. An annual average of just over 10 percent of the high-skilled and about 15 percent of the middle-skilled workers in the younger age group found new jobs in low-skilled occupations in the years following job displacement. In the lower skill group for which the labor market is even tighter, among those who found re-employment, 66–70 percent remained employed in low-skill occupations and only 3–8 percent moved up to middle- or high-skill positions [23]. These shares might be considerably different if labor market conditions were less robust.

4.5 Societal Impact of Job Displacement

The disruptive potential of technology creates deep concerns that society will become so split into winners and losers, or have-nots and the haves, that it won't be stable. Is using AI to upgrade work only a moral issue? Or is it also essential to invest in enabling broader, more distributed prosperity? Economics provides a cautionary answer. The "curse" of diminishing median income from these kinds of massive displacements is a deep, dark precedent from the first industrial revolution. But are companies or even society, in the tradition of past New Deals, being exhorted by the dreamers promoting AI-enabled robotics to inherit a portion of the economic blessings these technologies are likely to actualize? Or is it the case that, without more investment in processes that reinstate some of the stricken and their families at a respectable societal level, the consequences could be dire [22]? The experience of the '20 s and the '30 s clearly suggests that technology on the table is not enough. The history lesson is salient too: Startled by the fallacies and frauds that emerged as their countries industrialized three decades earlier, many European countries proudly signed on in the following thirty years to policies of nationalizing property and annihilation of competition in the market system. The deep-seated popularity of certain ideologies and the force of political movements supporting economic totalitarianism is evidence of the despair that such jobless societies have birthed. Although it is tempting to believe that workers displaced by AI and robots can transition to creative or service jobs at a slightly more accelerated rate than the time constants associated with previous technologically driven work have been, these are fundamentally different from the jobs that defeated AI [20].

4.5.1 Psychological Effects on Workers

Even if the purveyors of AI offer projections that jobs will not quickly disappear, discussions often steer cultural pessimists toward the probable negative psychological effects on working people. Researchers report that the intense pace and volatile nature of work in industries repeatedly prone to reorganization causes many employees to despair from chronic insecurity and powerlessness at work. In a more current and relevant finding, it is reported that when companies have already introduced changes to work practices in the name of greater efficiency, it is as though the purpose of work is being redefined away from its core meaning. These initiatives obfuscate work with so many paradoxes: it is fulfilling but frustrating, meaningful but meaningless, satisfying but unsatisfactory [6]. Many employees who have deskilled jobs feel that they have lost "proprietary" knowledge that others cannot replicate, despite what they realize as one-dimensional experts or algorithms sprinting through detailed job segments of tasks over or next best action. Such ploys for efficiency often feel like big practical jokes taken to the extreme, where managers claim that optimizing productivity is fun. Implementing new employee monitoring systems or other technologies in hidden service encounters or production processes intended to be cost-effective and transparent can deepen employees' sense of anger, isolation, and vulnerability. Employees also feel thwarted when customers have unfair advantages from being demanding, critical, and hostile, or abusive and hostile, because managers care only for many deeply dissatisfied consumers [24]. Frequent sales calls and worries about making performance targets have been said to increase nervousness and depression for many fast-food or retail employees.

4.5.2 Shifts in Workforce Demographics

As advanced economies are aging, an increasing percentage of workers are approaching retirement age, which could reduce the labor force and potentially increase the ratio of retirees to the workforce. With lower-income workers supporting a greater number of retirees, a decrease in the labor force would likely result in increased occupational stress and public sector outlays for social welfare programs. While the pending demographic crisis is well known, policymakers and economists have largely ignored its implications. The second great demographic shift underway is the aging of developed economies. At the beginning of the twenty-first century, the labor force growth rate began to decline in the United States, in large part because the baby bust generation was substantially smaller than the baby boom generation, which came of age as our workforce fifty years later [25]. In 2010, this phenomenon became a trend with speed and scope unprecedented in history. Western Europe's working population is projected to fall by up to 40 percent in the next fifty years, Japan's by 50 percent. As aging populations retire, economies will have fewer workers while the number of pension recipients escalates. In Italy, for every working person, there will be three pensioners by 2050. America's balance will be a little different, with three people of working age for every pensioner. The balance is already becoming more costly and complex [26]. In 1950, Japan had 12 people of working age supporting each retiree's benefits and supporting humanity, as presented in Fig. 4.2.

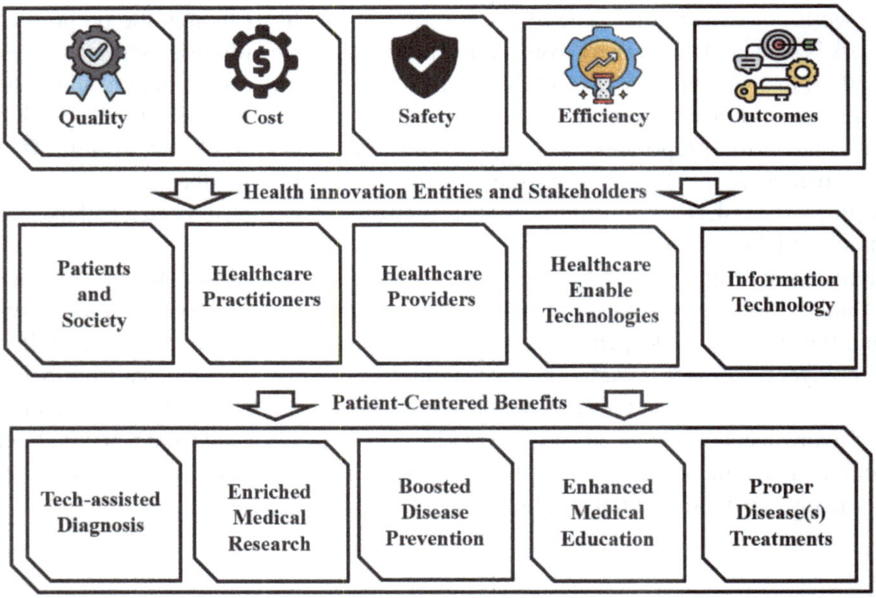

Fig. 4.2 Patient-centered benefits of artificial intelligence

4.6 Policy Responses to AI and Job Displacement

Government and corporations should collaborate on a National AI Strategy and other specific policy development aimed at addressing the AI transformation of the labor market and its economic impacts. A policy is needed to promote workers as shareholders and institute changes in corporate governance and labor organizations. Global taxation and corporate governance reform should be supported by policy. Markups rose during 2000–2017 as capital's share rose relative to labor's [27]. The power of intellectual property to support high corporate profits and rapid technological change, largely due to AI, must be addressed. Egalitarian ideologies of tech company founders and management need to give way to the existence and needs of vast classes of substitute workers. Reforms are needed to tax global corporate profits at the maximum federal statutory rate, and when that point is reached, Congress should reform laws to encourage repatriation of corporate profits without eroding tax revenue dedicated to the public sector and directly engaged with growing income inequality [5].

A National AI Strategy is needed to lead America into the 2020s. The United States leads the world in AI, and much can be done to ensure that American workers share in AI-generated productivity growth. Industry has taken the lead, but its fragmented approach is prone to long-term failure and short-term advocates of shareholders rather than stakeholders. The effects are being felt daily by American workers and will soon be felt at the federal Treasury. In a joined-up manner, the nation needs to create tax policy, labor market institutions, and corporate governance reform designed to keep America at the forefront of AI research and labor market gains. Technology on its current path could outstrip the ability of labor markets and existing institutions to deliver benefits and security to workers while raising incomes for the country [28]. The threats of accelerated inequality and wage suppression are real and need to be addressed.

4.6.1 Universal Basic Income

There is a spectrum of views on the type and extent of impact of the future of work. Two opposite views are the 'deflationists' and the 'full automationists'. While the deflationists argue that AI will not have a significant negative impact on employment, with new jobs and job substitution modalities arising to replace those jobs that will no longer exist, the full automationists argue that automation will result in mass unemployment and an accelerating mismatch between obsolete capital and labor assets and evolving new growth opportunities. Somewhere in between these two extremes is the growing discourse on Universal Basic Income (UBI). UBI is where every citizen receives a basic income paid monthly, regardless of whether they work. People receive the UBI irrespective of their means, who they are, and what they do, thus implying the creation of a direct link between work and income [29]. With UBI, it does not matter whether you are disabled or not, or whether you run a multinational company or are unemployed; everyone gets the same amount every month from the state as a legal right.

Understandably, the topic of UBI is highly contentious. This is because the level and value of work and the appropriate incentive effects of work are controversial and highly contested. Advocates might think that UBI is a way of modernizing the welfare state and simplifying it, freeing workers from what they view as boring, demeaning, and laborious tasks of remunerative work. UBI is also viewed as a way of supplementing the income of workers who have more than one job but whose take-home pay is still insufficient to meet their outgoings and obligations. On the other hand, opponents might call UBI payments 'moonshine' as these will entail a significant fiscal burden [30]. The leading industrialized countries have the capacity to finance UBI, but 'moonshine' supporters believe that UBI will be a massive transfer of resources from the middle class and core workers, who are the backbone of economic and political stability, to a declining low-wage and low-skill stratum of society. Albeit the welfare state is far from perfect in solving economic and social problems and technological unemployment is often seen as an economic myth, the demand for the abolition of fragmented welfare provisions and their replacement by a harmonized and more efficient UBI policy takes a rose-tinted view of the intrinsic nature of the contemporary labor market and the efficacy of UBI in ameliorating the perceived iniquities of contemporary capitalism [31].

4.6.2 Job Retraining Programs

As an alternative to infrastructure spending to rapidly pull us out of a job shortfall, many economists argue for a stronger commitment to training programs to help older workers who have lost jobs in traditional industries or who fear loss of jobs in other industries due to globalization and technological advances. Indeed, such programs have seen considerable success in Scandinavian countries. Five to ten years seems a reasonable period in which to retrain older workers for new careers before they lose patience and resort to unemployment benefits, deterrence via replacement labor. However, it is extremely difficult to find good jobs for people who have spent their lives in an industry that is now losing jobs due to globalization or technological advances [32]. One recent proposal comes from a leading figure in the industry most responsible for unemployment, who has proposed that robot manufacturers and others that are gaining jobs be taxed so that other workers can be better educated. This proposal is similar to taxing winners, and it is a highly unusual perspective. I favor this step simply because we cannot continue the rate of foreclosures, uncollected bills, and consumer bankruptcies known today without serious setbacks to the notion of global economic stability. The problem with creating jobs rather than training older workers to take them is that jobs have a hard time being created to match the skills of workers entering an alien discipline, especially if the discipline requires skills in math or statistics that the economy is used to finding only in recent graduates, such as actuarial or company statisticians [33].

4.7 The Role of Education in the Future Workforce

Critics indignant about the public schools' inability to teach young people coding or about every school district that fails to keep up with the fast pace of incorporating new technologies into the curriculum are misunderstanding the real challenge and the real role of education. In fact, the more successful educational system will do more than teach technical skills. It is important that those who are now advocating that at least some focus on character, discipline, and resilience be given greater weight in an expanded public school mission are the same people who, twenty years ago, would have declared the very same goals as too soft. And that goes for a premium on critical thinking and analysis that promotes a questioning spirit within a framework of knowledge required as a requisite to engage in that discussion that accompanies the knowledge or arises from that framework's lens on the world. Social mobility requires more than technical skills [34]. Most wage earners, regardless of how well-trained they might be, will need to function in a society in which a full life requires more than bread and circuses, and that the development of other non-technical skills, the pursuit of other worthwhile goals that may not pay off in the marketplace, the acceptance of the equality of every individual on terms that are not able to be priced on an income scale, are a necessary prerequisite to sustain viable democracy or make life worth living for its own sake. These goals give critically important insight into what our economic and workforce requirements must be in the distant future. The role of education, and of educational diversity, is far greater than being the source of these labels for individual motivations [35]. It is the prerequisite from which the generation of capital, if not prosperity, flows.

4.7.1 Adapting Curriculum for AI Skills

Educational institutions are realizing that they need to train students to work alongside AI. However, the challenge to integrate AI knowledge into academic institutions offering diverse curriculums and degrees is significant. Recently, a significant effort has concentrated on the need to move education into the digital age. The report suggests supporting the formal educational reforms that will move education into the digital age and at all levels: K-12, vocational and trade schools, higher education, and lifelong learning [36]. The report even highlights key elements of such forms: enabling digital education for digital careers, moving to mastery-based learning with flexibility to promote natural abilities and facilitate the attainment of transversal skills such as creativity, collaboration, and problem-solving [37]. Brazil believes that current capacity-building efforts are not preparing the workforce at all for the advent of the age of automation in production. Research being conducted proposes integrating different subjects, which are organized as pedagogical experimentation, an open teaching–learning assessment, carried out in detecting the natural interaction of space–time and artistic expression of different subjects. This is shown by the development of innovative pedagogical practices aimed at fostering new content, specific lexicons, and thus specific results. The great potential of this work is its ability to embrace many subjects and educational segments, encouraging different levels and helping the teaching–learning processes [26].

4.7.2 Lifelong Learning Initiatives

One obvious policy response to the AI jobs challenge is lifelong learning initiatives. It includes training programs in data science, machine learning, and artificial intelligence for employees, customers, and the public. There is a focus on providing opportunities for students and working professionals in underserved regions around the world. There has long been a promotion of computer science, and numerous tools provided for that community. But the magnitude and future challenges we face will take a far greater collaboration. Critics argue that such programs will only help a few. Interestingly, recent research says the jury is still out as to whether upskilling the workforce will really close the wage gap between workers able to somehow stay ahead of AI versus those who can't. Preliminary program assessments of these initiatives have shown mixed results [38]. Optimistic and pessimistic commentators on lifelong learning agree on the importance of getting the workforce ready for the new AI technologies.

Pessimists worry about how to scale and fund the programs from both private and public perspectives. Optimists believe the programs could help smooth the transition to the future world of work. One lesson from other countries is that U.S. students don't have to make the transition to the future of work with such a high reliance on college education. A premium in wages and benefits is still a valid investment even after inflation; it is simply higher than it ought to be. Since a BA certainly has value, the question is can we come up with alternative efficacious and less costly ways to achieve the same positive result. A study found that apprenticeship programs created a good match between the skills bestowed by the program and the hard-to-quantify soft skills that make it far easier for a worker to fit into the manufacturing industry culture [39]. It is argued that American workers are not radically different from their European colleagues and that a similar apprenticeship-based training program could appeal to many American students.

4.8 Ethical Considerations in AI Development

Ensuring that AI during any industrial revolution levels the playing field for everyone—and does not merely seize the potential, profits, and control by the few for the few—involves challenging and potentially transformative work. This work could include addressing these questions: Who "develops" AI? Who determines who the AI serves and for whose benefit? What principles act as guidelines when making tough decisions about AI choices? These may not be easily solvable questions, but they are critical to achieving a responsible and ethical AI future. Multiple sectors should play a role in deciding the principles that guide various individual industries' AI development. These sectors could include ethics experts, human rights organizations, corporate culture experts, customer groups, unions, and worker representatives [40]. The promise in AI and digitized technologies is using them to enhance humanity, not diminish it. To fulfill this potential, the private industrial sector has an important ethical obligation to people and governments to actively develop and adhere to ethical guidelines and increasingly strict industry standards that, at a minimum, prevent AI from infringing on privacy, silencing voices, restricting rights, or causing more harm than good. While private interests may develop AI most effectively, they must plow profits and know-how back into initiatives that serve the uplift of society by providing opportunities for those who are at risk of being left behind by AI's progress. All sectors of civil society, not merely the business sector, have a role to play if a responsible future of AI is going to be realized [41]. With so much at stake, we must work together to ensure that the future of work is as bright for all as AI's evolving potential allows.

4.8.1 Bias in AI Systems

One question that can hamper labor market policies is whether we can trust AI systems to make consequential employment decisions fairly and accurately, especially given the evidence of their often substantial biases. This is particularly relevant as many of the decisions made by AI systems can be considered consequential employment decisions. This is true on the candidate side, as these systems often decide who gets to be interviewed or who gets the job. This is also true on the employer side as AI systems help make decisions related to productivity and team composition [42]. However, as we have seen, especially AI hiring decision systems can easily replicate and further amplify discrimination against minorities and women, often leading to severe concerns for not only those individuals who are being disadvantaged but for society as a whole. Therefore, the capability of AI systems to be used to aid in such decisions also creates the imperative to build them in such a way that they help reduce rather than amplify existing sources of bias. It calls us to question when AI system decisions can be considered fair or just, and under which circumstances they can be trusted to make such decisions automatically, if ever. Additionally, it also calls for societal institutions to enforce clear guidance on how these systems are allowed and not allowed to make decisions, what kind of data they should be trained on, the bounds they should be kept within, and, if violated, the possible and severe consequences to foster their development in such a way that they actively support ethical and unbiased considerations of human resources [43].

4.8.2 Accountability in AI Decisions

While the need to table the wider category of accountability has been clear from the beginning, it may be possible to take a gradual approach to machine accountability. The idea of the gradual approach is that in the early stages of developing AIs that will play important roles in human lives, we create AIs that carry personal accountability for their actions. Such personal accountability is imposed or maintained by whatever controls the AI; it need not be real personal accountability. In the case of AIs displacing humans in the broader political-economic space—voting and working—the most prominent human attribute that will be needed is empathy. It may be a mistake to think of assigning AIs personal accountability for their decisions rather than their creators [44]. The result of decision accountability for the political-implementing AIs is rather striking. Acting political, empowered AIs will be in the position of deciding who can vote in elections, who can have robots, and all of the other political questions traditionally decided by democracy. In limiting the physical autonomy of humans in the economic sense, AIs will become the political darlings of the elites, who always disliked democratic constraints—if this kind of elite ever tolerated them. The important part of this paper's central argument is that questions about work and AI are questions about political economy. We are putting our faith in what amounts to two dangerous delusions—as once remarked, no country has long been ruled by its philosophers, or if we believe in current democratic government systems, our administrators [45].

4.9 Case Studies of AI Integration in the Workplace

Overall, these case studies were selected to underline the diversity of work tasks, AI technologies, methods of task displacement, and outcomes from AI intervention. However, in several cases, given the context, the presence of AI is more implicit than explicit. Whether AI is a possible or actual threat to work stability depends at the very least on the presence, nature, and dynamics of interactions resulting from AI functionality, and these case studies provide 10 settings with the potential for a wide variety of patterns. The following chapter will argue that considering work in sociotechnical terms can provide a better understanding of potential implications. AI's arrival into work tasks is often announced with shiny, bright corporate panels showing the latest successes from early adopter technology companies. My examples are certainly not representative of the general AI-using situation, the vast majority of which is unknown or hidden. They are therefore potentially biased [46]. We, however, make no specific claim for the baseline frequency of these systems, nor for the representative level of their engagements with associated workers. These examples are heterogeneous and do not have a straightforward summative production story, as presented in Table 4.1.

4.9.1 Successful AI Implementations

Many companies—large and small—have successfully implemented AI solutions. Typically, these companies followed a simple, orderly process in which AI was used to augment human activity, not replace employees. The AI applications discovered new customers, recommended upsell/cross-sell additional products to known customers, or fine-tuned overall service and quality offerings. Every time the applications worked, new products were defined and efficiency improvements derived from the use of AI were distributed to some combination of customers, employees, or shareholders [47]. At their best, it was a win–win set of scenarios that showed the transformative possibilities of AI. The rule of thumb for a successful AI tool in the near future will be that the tool can solve a clearly defined, limited, and specialized test. For some problems, this rule means that AI is guaranteed to fail: a tool could not be built, and thus the problem represented a grand challenge in artificial intelligence. Build a prototype of the AI tool, assess whether it effectively solves a business problem, and then decide whether to deploy it, often in a decision-support role, to augment employees and customers. After AI tool deployment, always closely monitor whether the tool provides benefits or needs retuning [48]. With this approach, AI-enhanced capabilities make business operations better, providing rewards for those who pay for the development and implementation of AI-enabled means.

4.9.2 Failures and Lessons Learned

We have now dug into many aspects of the different projects and our view of them as case studies in success and failure. In the section below, we review what can be learned from these deployments and look at commonalities between projects to identify potential contributing factors to success or failure. Since there is not necessarily a baseline with which to compare projects, what constitutes a "failure"? This is an important consideration, since some projects may seem in fact to be successful. It is unclear where the boundary lies between success and failure for some projects. A rule of thumb might be: if the project solves the problem (and hence does so in a socially beneficial way), it could be a success. Conversely, if the project either does not solve the problem or does so in a socially harmful way, it has failed. For instance, a mismanaged security deposit did not solve the problem but

Table 4.1 Artificial intelligence's threat to jobs and economic stability

Application	AI technology	Job impact	Economic implication	Affected workforce	Suggested solution
Manufacturing	Robotics and automation	Mass reduction in manual labor roles	Regional unemployment and skill gaps	Factory workers	Invest in vocational training, smart factories
Retail	Self-checkout systems	Fewer cashier and sales assistant jobs	Loss of entry-level jobs	Retail staff, youth workers	Customer experience-focused reskilling
Transportation	Autonomous vehicles	Threat to drivers (truck, taxi)	Reduced income for gig economy workers	Drivers, delivery personnel	Regulation, reallocation into logistics roles
Finance	Algorithmic trading	Fewer analysts, clerical staff	Job polarization, high-income concentration	Mid-level professionals	Ethical AI practices, data literacy training
Healthcare	AI diagnostics	Reduced demand for some diagnostic roles	Efficiency gains, but potential job shifts	Radiologists, technicians	Human-AI collaboration frameworks
Agriculture	Precision farming	Reduced labor needs on farms	Rural job losses, tech divide	Farm workers	Tech access for rural communities
Education	AI tutors and grading	Threat to teaching assistants	Standardized education, potential job shrink	Tutors, adjunct staff	Redefine teaching roles with tech integration
Customer service	AI Chatbots	Call center downsizing	Cheaper operations, high turnover	Call center agents	Emotional intelligence training for escalation
Legal sector	Document automation	Fewer paralegal and junior roles	Lower costs, fewer entry points	Junior legal staff	Upskill in legal tech and compliance roles
Journalism	AI content generation	Replacement in reporting and editing	Content overload, decline in investigative journalism	Writers, editors	Emphasize creative and investigative journalism

was priced into the loan accordingly (and hence "solved" the problem for all parties with appropriate compensation) [49]. If, on the other hand, solvency was incorrectly decided and unsuitable credit was offered, this had a negative social impact (with added risks). While this is not a perfect measure, it is at least a simple measure to decide between success and failure.

4.9.3 Future Scenarios: Optimistic vs. Pessimistic Views

The debate about the future of work often revolves around two main scenarios. The "optimistic" view proposes that AI and robotics will enable unprecedented advancements in human well-being, enriched by new products and processes and untethered from the time-consuming requirement of having to work for a living. The "pessimistic" scenario foresees mass irrelevance and mass unemployment as the vanguard of technological unemployment takes work away from vast parts of the economy, leaving few employment opportunities for workers not skilled in the tasks that machines can't do [50]. Economists refer to these two positions as the "substitution effect" (tech will mostly displace workers and create deep problems for our skills-based economic system) and "income effect" (the newly created economic activity and the benefits of that activity will raise incomes and create job-creating spending, wealth, and societal well-being) of technological progress. The task for both schools of thought is to outline what kinds of economic policies would best channel technological progress and its benefits in a way that promotes the optimists' positive scenario. The income effect path is the desired route, one that can help preserve a demand for workers at all skill levels [15]. In this, common ground must be found, as no modern economy can afford mass unemployment, nor can it prosper without an adequate and growing demand for products and, thus, workers.

4.9.4 AI as a Job Creator

Contrary to the doom and gloom that seems to be typical of discussions over robots eliminating jobs and artificial intelligence pushing people into the unemployment lines, there is a quite viable and much more encouraging scenario; one that is supported not only by common sense, but by careful analysis as well. And that is, in the long run, artificial intelligence can become a major job creator; one that lifts all of society and not just the corporate elite. It can also lead to desirable major changes in the nature of work. However, for this favorable scenario to materialize, society will have to make a series of major decisions that point to attitudes that are entirely the opposite of those that are now prevalent [42]. One must note well that research progresses by small steps and that major achievements are built upon many such steps. So too, will be the result of the accelerated stream of innovations that is now emerging from the computer-based revolution. And, when we analyze the effects of such transformations, we must keep in mind that each set of changes will be but a single step in a continuous process. Both common sense and conventional theory tell us that it is large numbers of such sets of steps that cause important problems. We will soon be in a position to assist; both trained and unskilled workers will become profoundly more effective [26].

4.9.5 AI Leading to Economic Inequality

On the other hand, AI could also lead to economic inequality. Organizations are hiring scientists, engineers, and mathematicians to leverage AI for competitive advantage. In cities like Silicon Valley, unemployment is the lowest. In the large swath of mid-America, where traditional industries are declining, economic prospects are poor. As the gulf between artificial intelligence "haves" and "have-nots" widens, there is the potential for economic inequality spreading to income and wealth inequality. There's no certainty that income and wealth will become more unequal [8]. Other factors such as increasing public recognition of AI's threats to jobs and economic stability and policies that redress the imbalance may prevent economic inequality from widening. Over the long term, and combined with other advances in robotics, genetic engineering, and nanotechnology, AI may assist in the dismantling of the current economic system. It's not so much that AI is fundamentally anti-democratic or evil; it's that arriving at the right balance is a difficult, yet important, economic, political, and ethical task [3].

4.10 Expert Opinions and Predictions

Many experts presented enlightening and sometimes discouraging opinions and predictions on this rapid advancement in AI technology and the evolving future of work. It was widely agreed that AI could be a device of immense positive value, enriching the lives of countless people who are currently suffering and toiling under the weight of human conflict and lack of resources. AI would enable them to bypass the unfair competition favored in the future. Instead, people would spend much more of their time thinking about improving the natural world, solving some of its most fundamental instabilities and injustices. However, this could only happen as long as we had a good education. If we wanted an AI-enhanced economy, we had to determine a way to provide everyone with an education. Economic right-to-work policies were severely limited [28].

In capitalist societies, they were not enough to guarantee that the average person would necessarily benefit from advances in AI. Moreover, any system that excluded large numbers of people was something of deep concern. We had to accumulate extraordinary resources to deal with the troubling future of tens of billions of people literally transformed out of their jobs. In the end, AI would supplant many jobs, and governments should support that prospect as much as possible. We had to increase defenses for the industries that were being destroyed and prepare a safety net for the people who succumbed to that enclosure. It would be great if we could rely on the smooth continuation of labor market dynamics to manage workers, but it wasn't very smart to enter AI without developing sure-fire alternative protections. It would be great if we could be confident that we would not be propelled into a future without work as the only ironically profitable company, but we needed to consider the alternative world where machines were all. At a time, the claims of the majority of people in the United States own that, and a somewhat luxurious time. Sooner or later, we were going to want to solve these difficult problems [51]. And, if we did not, someone else would.

4.10.1 Interviews with Industry Leaders

Siddharth Suri, an AI researcher, was asked what advice he would give someone pursuing a career in AI. It's an interesting question: are AI jobs at risk? Are there things prospective workers should look for in order to have some measure of income safety if AI does become a significant contributor to job losses? Suri said that jobs that have higher levels of flexibility, unpredictability, social interaction, and require complex value judgments will be in higher demand, at least for the foreseeable future. He also pointed out that writing human instructions for robots will be a growth area and that creative jobs will last longer with the arts as a valuable degree. Neil Jacobstein is the head AI guy at a university [30]. He must think about the future of work a lot, so I asked him what he thought. Jacobstein is philosophically very much on the same page: "Interest and core skills in AI and materials science are strong bets for the long term. Flexibility, unpredictability, and complex value judgments are attributes associated with jobs that will be less subject to AI replacement in the near term and mid-term future." His suggestions: "Major in visual arts, engineering, science, finance, or writing, and throw in philosophy." To the best of my knowledge, not a single philosopher has lost their job to automation, much less to robots. However, only about 140 students in the United States graduate with a Bachelor's degree in philosophy on average [33]. Maybe that should be, like, your backup major to computer science?

4.10.2 Predictions from Economists

There are predictions from experts on the impact of machine learning on labor and the economy. First of all, as we are pushing to use predictive analytics in broader circumstances, it is crucial to reflect on how these and similar tools may ultimately affect our workforces and economy. It is up to businesses, policymakers, and the public to decide whether to shift the benefits associated with machine learning back to profits, or if the best path forward includes considerations for workforces and, ultimately, the economy. A point is made that if machines are doing all the work, nobody has a job, so nobody has any money, so the economy rolls over. It is noted that we have forgotten the first link in the chain [26]. We have been told that the rise in profit is the result of technology, but it is increasingly clear that it is caused by much more mundane phenomena, such as the decline in union power and the policy of low wages followed by multinational corporations. The benefits of technology have been captured between the capitalists at the expense of the workers. Companies spend less on workers and engineers to make the robots, then boast that they have pushed the boundaries of what is possible [49]. The result is that people were reduced to the federal equivalent of horses in a certain movie.

4.11 Global Perspectives on AI and Employment

As more decision-making power and economic resources are concentrated in fewer people's hands, the willingness to increase income and wealth taxes has lessened, a trend that has, in turn, fed income inequality. Since workplaces need to be transformed to take full advantage of AI's capabilities, worker preferences, values, skills, and needs are planned to root economies that are entering a phase of growing instability. Transforming the workplace in this fashion could then prove to be a key tool for ensuring wider benefits to society as a whole derived from AI, and that a sense of fairness is restored to its technological and economic developments. Whether a state-based income policy, a more flexible labor market, or something entirely different is chosen, however, social considerations need to be prioritized in order to make AI maximally beneficial to the greatest number of people [23]. The largest societal impact of AI will be how economies and people intersect. The increases in productivity generated by AI are calculated to be sizable, with the boost to the US due to the latest AI techniques being estimated to total \$1.2 trillion, and the Chinese economy estimated to boost its GDP by 26% by 2030, or \$7 trillion, over the expected baseline. The potential to implement AI is expected to lead to the total economy impact having a global value of \$13 trillion in 2030, which is equivalent to 1.2% additional GDP growth per year. This will enhance GDP per capita in developed economies by 36% over the same period. The report calculates, using macroeconomic models, that renewed economic growth driven by AI can enable a range of socially beneficial uses, such as a decrease in the amount of time spent at work or further investment in non-market activities. To that end, the findings are that the largest near-term effects of AI will likely be in China, where AI is expected to contribute an additional 1.6% per year of GDP growth towards the end of the forecast period, and where widespread use of AI in autos and tech sectors would contribute the most to economic growth [45].

4.11.1 AI in Developed Countries

It's easy to be lulled into the complacency that, since AI is operating everywhere around us, the whole world has an AI economy. Takes a trip around the globe to less-developed countries and that complacency evaporates quickly. While AI systems have been demonstrated in almost any AI discipline—robots with AI neural networks that "think" and grasp objects; AI vehicles that navigate unknown environments; medical AI systems that diagnose diseases better than top human radiologists—these demonstrations have taken place only in the most highly capitalized AI laboratories in the most developed countries. The few AI technology companies that we do have keep adopting more AI technologies, keeping the gap between them and countries' AI and R&D capabilities huge—too huge for most countries to begin to catch up [48].

Despite this hype, AI's capabilities are narrow. AI is not a single known thing but rather is a collection of almost 20 technological paradigms revolving around a few core concepts such as reasoning, learning, semantics, and agency, which sensory modalities map to these core concepts and extra efforts have to be invested to map to more of them, and why. But why do this? AI has been used as a metaphoric concept for a long time, so long that too many tasks with the slightest connection to AI get called AI tasks and few notice that this custom domain and AI proximity connection is just technologically flavored handwaving. Advancing AI science into engineering and guiding such advances toward forms of steel that pay handsome global dividends remains a grand challenge, both in the science and in the technologies that are just at the modern scientific level [26]. Even so, the "front door" for AI to the world economy might profit from boosting the simplest, near-term vision tasks in our instances of AI's more elemental paradigms to practical capabilities in the near term.

4.11.2 AI in Developing Economies

The greatest impact of artificial intelligence (AI) on jobs and economic stability may come in the developing world. Just as countries in Asia and Africa were starting to climb the manufacturing job ladder, the ladder is being pulled out from under them. With current global trade tensions and rising nationalism, they are losing their foreign aid and may not be able to attract enough high technology industries to take up the slack. Recently, East Asian newly industrializing economies have taken over the climbing ladder. This historically new development enabled many countries to inject themselves into a growth process without an extended period of labor-intensive light manufactures, such as textiles, toys, and shoes. But AI raises the bar for technology industries. It is a great opportunity for nations with excellent computer and data skills to leapfrog forward [8]. However, for most countries, AI is likely to widen the difficulties arising from a changing employment environment, as the household divisions of labor are disrupted by the loss of jobs to machines in the more skilled routine cognitive areas.

4.12 Technological Innovations on the Horizon

The current pace of technological innovation is notable for its speed. In the realm of technology, especially, 'tomorrow' is indeed often today. The advances described up to now, significant as they are, only scratch the surface. Other innovations that may make even more significant changes in the social and economic order are currently being tested and developed. Once perfected, in twenty years or less, they may alter all phases of human life. Some of these developments are quite new, and the possibilities they present are just beginning to be recognized. In other instances, long-term research is in progress, and the potential of these fields has long been known. For decades, the laser was usable only as an interesting but largely irrelevant scientific curiosity before its current applications became apparent [4]. Similarly, since the invention of the computer, it has steadily spread from its original use in simply adding and subtracting for an atomic energy laboratory to its present industrial, commercial, and scientific applications.

Certain developments were not included in the 1970 survey; they had only recently received significant funds through military or space grants, and their possibilities for use at reasonable cost were beginning to be explored. In most cases, however, these 'new' fields already had a substantial amount of published work done through the laboratory stage. Today, there is a new group of potential technological innovations in progress. Some, and indeed some entire fields, were not even identified, much less supported, until a handful of scientists realized the significance of achievements in other unrelated fields. Many of these mentioned would not be considered, even today, by most citizens, unless they were already involved in these areas [5]. Examples include integrated optical circuits, bio-ionics, and satellite remote-controlled sensor structures. Some of these brighter stars on the horizon—hopefully open to all cultural, technological, or economically advanced nations and peoples—will be mentioned. Their potential could be achieved by the joint development of more than one country or people, a process started and furthered by such natural pioneers and inventors who may have already begun collaborative work with distant friends or colleagues. Such a trend, if nourished, could unite otherwise isolated regions or individuals. With their essential common interests, it might advance those 'other activities' discussed. Furthermore, these synergistic people would have assigned even more of their resources for the common betterment of humanity [8].

4.12.1 Emerging Technologies

The two most important broad trends about work in the current century are the rapid advancement and diffusion of digital technologies, which are making firms more productive and their workers wealthier, and the increased pressure on wages and job stability at the middle and bottom of the income distribution. While it would also seem obvious that cures for workers' concerns have been around for over a century: more education for workers and their children and more retraining, as more and more skill development and reorganization is inevitable, difficult, costly, and risky to all concerned. Education and retraining policies that help ease the transition to digitized work can mitigate the negative labor market pressures that have arisen thus far. However, the need to overcome political resistance does not inspire confidence; not a single developed economy has been able to systematically reform its labor market institutions to deal with the recent morose labor markets [28]. The technologies that are often discussed include parallel processing, distributed ledgers, and 'true' artificial intelligence that is, machines with the ability to learn from their experiences. These technologies differ from previous innovations in that they combine the analytic capabilities of old technologies with improved mechanical ones, and they can 'sense,' 'predict,' and 'respond' to an evolving set of circumstances without any human input. Since 2009, digital technologies have gone through a period of rapid experimental and incremental improvement [32]. In the procedure system infrastructure space, we have seen the rise of massive parallel processing architectures and, from them, the distributed ledgers that underpin digital currencies.

4.12.2 Future AI Capabilities

We now discuss two issues that are critical to the role of AI in labor: understanding the limitations of current AI compared with future, more generalized AI; and understanding the interplay of AI's technical capabilities with the longer-term economic dynamics underlying wage inequality. The failure to reflect the limitations of current AI—due to overhype around AI—may have led a previous generation to be in a permanent, rather than temporary, slack adjustment. The lack of appropriate understanding of longer-term economic dynamics makes recommendations for public policy—with an AI-specific focus or otherwise—with respect to labor outcomes unclear and decidedly conflicting. Much of the media discussion around AI generates hype that is not grounded in evidence. It is critical to appreciate what AI currently can and cannot do, the ways in which it is now a unique capability compared to previous technology, and the unwise claims [37]. AI, and in particular, recent AI, is the current term for automation. Automation is a technology-driven shift in the composition of production. Automation has occurred for over 200 years, and thus many sectors that formed the employment base of any developed country's economy a couple of centuries ago are extinct. It is well understood. AI is unique in that it can automate these tasks in offices as opposed to on the farm or in the factory. However, this isn't a superpower: most work is not conducted in neatly circumscribed offices at desks with computers that are easily able to perform the kind of automation that generates predictions about the wholesale elimination of people from the office environment. AI's effects in the retail and manufacturing sectors are also limited. There is a widespread belief that AI might actually be at its peak effect in these three sectors [45]. However, to the extent it can be unshackled, the biggest impact of AI will be in healthcare, where there are very few income-constraining effects of automation that are economically benign.

4.13 Conclusion

To fully prepare workers for changes in the labor market due to AI, policymakers need to anticipate the future's demands and make the U.S. labor market more adaptable, efficient, and resilient, among other developed nations. To do this, they can adopt policies, regulations, and programs that aim to support collective bargaining, update the social safety net to meet the demands of current and future workers, invest in education and worker training, and change regulations. Additionally, some public labor market policies should only be undertaken if they are sufficiently complementary to the market's normal operations or if policy interventions can realistically address any adverse consequences. If technological advances result in substantial dislocations in the labor market, the labor market's overall adjustment mechanisms, such as wages or hours worked, may behave differently than in the past. Policymakers should be careful about interfering with these mechanisms unless clearly justified. Technology has been changing jobs and tasks for thousands of years, but recent advances have raised concerns that automation, and in particular artificial intelligence, will prove exceptional. This brief has provided an overview of some of the issues policymakers and economists are coping with today, including the outlook for future improvement and AI's potential effects on the future of work and productivity. In summary, while there is broad agreement that AI will substantially change the nature of most jobs and will also create new kinds of jobs yet to be invented, the outlook for the broader trends in employment is less clear. There is considerable debate and uncertainty about the long-run effects of AI on productivity, how automation may be different from past episodes, and the potential for more profound effects on the distribution of incomes between labor and capital. It is this uncertainty and the inherent difficulty of predicting the future in a rapidly evolving field that present the primary obstacles to informed, robust policy.

References

1. Shafik W. Generative artificial intelligence for social good and sustainable development. In: Generative AI: disruptive technologies for innovative applications. 2025. p. 153–85.
2. Vaishali A. What is artificial intelligence? How does AI work, applications and future? Great Learn. 2021.
3. Wong RY, Madaio MA, Merrill N. Seeing like a toolkit: how toolkits envision the work of AI ethics. Proc ACM Hum Comput Interact. 2023;7(CSCW1).
4. Ng DTK, Leung JKL, Chu SKW, Qiao MS. Conceptualizing AI literacy: An exploratory review. Comput. Educ. Artif. Intell. 2021;2.
5. Moldt JA, Festl-Wietek T, Madany Mamlouk A, Nieselt K, Fuhl W, Herrmann-Werner A. Chatbots for future docs: exploring medical students' attitudes and knowledge towards artificial intelligence and medical chatbots. Med Educ Online. 2023;28(1).
6. Jain A, Ranjan S. Implications of emerging technologies on the future of work. IIMB Manag. Rev. 2020;32(4).
7. Paul S, Yuan L, Jain HK, Robert JR. LP, Spohrer J, Lifshitz-Assaf H. Intelligence augmentation: human factors in AI and future of work. AIS Trans. Hum.-Comp. Inter. 2022;14(3).
8. Su J, Zhong Y. Artificial Intelligence (AI) in early childhood education: curriculum design and future directions. Comput. Educ. Artif. Intell. 2022;3.
9. Shafik W. Generative adversarial networks: security, privacy, and ethical considerations. In: Generative artificial intelligence (AI) approaches for industrial applications. Springer;2025. p. 93–117.
10. Shafik W. Generative AI for social good and sustainable development. In: Generative AI: current trends and applications. Springer;2024. p. 185–217.
11. Shafik W. Introduction to ChatGPT. In: Advanced applications of generative AI and natural language processing models. 2023.
12. Shafik W. Deep learning impacts in the field of artificial intelligence. In: Deep learning concepts in operations research [Internet]. New York: Auerbach Publications;2024. p. 9–26. https://doi.org/10.1201/9781003433309-2
13. Shafik W. Toward a more ethical future of artificial intelligence and data science. In: The ethical frontier of AI and data analysis [Internet]. IGI Global;2024. p. 362–88. https://doi.org/10.4018/979-8-3693-2964-1.ch022
14. Shafik W. Ethical and legal considerations in artificial intelligence. In: AI-enabled threat intelligence and cyber risk assessment. 2025;p. 90.
15. Bankins S, Formosa P. The ethical implications of artificial intelligence (AI) for meaningful work. J. Bus. Ethics. 2023;185(4).
16. Wang D, Weisz JD, Muller M, Ram P, Geyer W, Dugan C, et al. Human-AI collaboration in data science: Exploring data scientists' perceptions of automated AI. Proc ACM Hum Comput Interact. 2019;3(CSCW).
17. Shafik W. Navigating emerging challenges in robotics and artificial intelligence in Africa. In: Examining the rapid advance of digital technology in Africa [Internet]. IGI Global; 2024. p. 126–46. https://doi.org/10.4018/978-1-6684-9962-7.ch007
18. Shafik W. Human-computer interaction (HCI) technologies in socially-enabled artificial intelligence. In: Future of digital technology and AI in social sectors. IGI Global;2025. p. 121–50.
19. Banafa A. AI and the future of work. In: Transformative AI. 2024.
20. Denno P. Cognitive work in future manufacturing systems: human-centered AI for joint work with models. J. Integr. Des. Proc. Sci. 2024;27
21. Kiwanuka F, Shafik W. Nursing as a lingua franca for artificial intelligence in patients' care trajectories: where are we headed? Nurs Res. 2025;74(3):170.
22. Faishal M, Mathew S, Neikha K, Pusa K, Zhimomi T. The future of work: AI, automation, and the changing dynamics of developed economies. World J. Adv. Res. Rev. 2023;18(3)
23. Jarrahi MH. Artificial intelligence and the future of work: human-AI symbiosis in organizational decision making. Bus Horiz. 2018;61(4).

24. Zhao L, Zhu D, Shafik W, Matinkhah SM, Ahmad Z, Sharif L, et al. Artificial intelligence analysis in cyber domain: a review. Int. J. Distrib. Sensor Netw. 2022;18
25. Shafik W, Kalinaki K, Fahim KE, Adam M. Safeguarding data privacy and security in federated learning systems. In: Federated deep learning for healthcare [Internet]. Boca Raton: CRC Press;2024. p. 170–90. https://doi.org/10.1201/9781032694870-13
26. Mahmud B, Hong G, Fong B. A study of human-AI symbiosis for creative work: recent developments and future directions in deep learning. ACM Trans. Multimedia Comput. Commun. Appl. 2023;20(2).
27. Shafik W. Leveraging emerging technologies for smart and secure software development: blockchain, IoT, and beyond. In: Modern insights on smart and secure software development. 2025;p. 41–76.
28. Frey CB, Osborne M. Generative AI and the future of work: a reappraisal. Brown J. World Aff. 2023;30(1).
29. Rožman M, Tominc P, Vrečko I. Building skills for the future of work: students' perspectives on emerging jobs in the Data and AI Cluster through artificial intelligence in education. Environ. Soc. Psychol. 2023;8(2).
30. Nyholm S, Rüther M. Meaning in life in AI ethics—some trends and perspectives. Philos Technol. 2023;36(2).
31. Bughin J, Seong J, Manyika J, Joshi R. The future of work in the era of AI. Boston Consulting Group. https://www.bcg.com/publications/2021/impact-of-new-technologies-on-jobs. 2021.
32. Fügener A, Grahl J, Gupta A, Ketter W. Collaboration and delegation between humans and AI: an experimental investigation of the future of work. SSRN Electron. J. 2019.
33. Howard J. Artificial intelligence: implications for the future of work. Am. J. Indust. Med. 2019;62
34. Lloyd C, Payne J. Rethinking country effects: robotics, AI and work futures in Norway and the UK. New Technol Work Employ. 2019;34(3).
35. Furendal M, Jebari K. The future of work: augmentation or stunting? Philos Technol. 2023;36(2).
36. Shafik W. Artificial Intelligence-enabled cybersecurity and internet of things applications in smart cities. In: Building tomorrow's smart cities with 6G infrastructure technology. IGI Global Scientific Publishing;2025. p. 301–34.
37. Róbert P. Will robots take our jobs? Inform. Tarsadalom. 2022;22(3).
38. Ellingrud K, Saurabh KE, Gurneet S, Dandona S, Madgavkar A, Chui M, et al. Generative AI and the future of work in America. McKinsey Global Insitute. 2023.
39. Stahl A. How AI will impact the future of work and life. Forbes. 2021.
40. Manyika J, Sneader K. AI, automation, and the future of work: ten things to solve for. McKinsey Global Institute. 2018.
41. Thakkar D, Kumar N, Sambasivan N. Towards an AI-powered future that works for vocational workers. In: Conference on human factors in computing systems—proceedings. 2020.
42. McKinsey Global Institute. Digitization, AI, and the future of work: imperatives for Europe. In: Proceedings of the European union tallinn digital summit September 2017. 2017;(3).
43. Carstensen T, Ganz K. Gendered AI: German news media discourse on the future of work. AI Soc. 2023.
44. Anthony C, Bechky BA, Fayard AL. "Collaborating" with AI: taking a system view to explore the future of work. Organ. Sci. 2023;34(5).
45. Tsiakas K, Murray-Rust D. Using human-in-the-loop and explainable AI to envisage new future work practices. In: ACM International conference proceeding series. 2022.
46. West DM. The future of work: robots, AI, and automation. 2018.
47. Liu J, Xu X (Cedric), Li Y, Tan Y. "Generate" the future of work through AI: empirical evidence from online labor markets. SSRN Electron. J. 2023.
48. Hammer A, Karmakar S. Automation, AI and the future of work in India. Employee Relat. 2021;43(6).
49. Lalioti V. RethinkAI (TM): designing the impact of AI in the future of work. In: Proceedings of the European conference on the impact of artificial intelligence and robotics (Eciair 2019). 2019.

50. Su Z, He L, Jariwala SP, Zheng K, Chen Y. What is your envisioned future? Toward human-AI enrichment in data work of asthma care. In: Proceedings of ACM Human Computing Interaction. 2022. p. 6(CSCW2).
51. Mariani MM, Machado I, Magrelli V, Dwivedi YK. Artificial intelligence in innovation research: a systematic review, conceptual framework, and future research directions. Technovation 2023;122.

Chapter 5
AI in Warfare and Security: The Rise of Autonomous Weapons and Global Threats

5.1 Introduction

The concept of machines fighting or outperforming humans has gripped the imagination, or more likely, fear, of writers, philosophers, and warriors for thousands of years. European knights of yore were equally enamored, or concerned, and by 1495, a famed artist and engineer created a rough sketch of a robotic knight, as well as drawings of machines that would work automatically [1]. Real machines and tools were to be some 200 years or more in the future before this vision would be realized, in part. The first use of truly autonomous devices, and with a defined purpose that was threatening to animals and humans, was with the advent of mines and torpedoes, late in the American Civil War. They were spheres filled with explosives that had synchronized large, sharp, and curved knives that were intended to thwart any attempts at disarming the device [2]. Once a target vehicle or animal touched the sphere, it ignited a fulminating and explosive powder to cause the primary filling to explode and destroy the intended target when hit.

5.2 Types of Autonomous Weapons

First, we should understand the types of autonomous weapons. One answer is to use the schema provided by the Department of Defense: Remotely controlled force regulation weapons are under direct, real-time human control, and thus are not autonomous. Tightly controlled autonomous force regulation weapons are programmed to adhere to strict behavioral constraints, which are applied with a high level of determination and reliability, undermining some useful autonomy. Loosely constrained autonomous weapons are programmed to adhere to behavioral constraints up to a pre-specified confidence level and retain sufficient useful autonomy to adapt their behavior in a dynamic environment [3]. Reactive lethal

W. Shafik, *The Dark Side of AI*, https://doi.org/10.1007/978-3-032-09130-7_5

autonomous systems design themselves to be reactive, following a deterministic sequence of operations and possessing adaptive controllers that adjust to dynamic environmental changes specific to the operational environment. A second answer is to focus on two dimensions of the weapons function: what is the degree of human involvement in decision-making, and how complex is the weapons' functioning? Concentrating here upon the autonomous weapons aspect of both the definition and the two dimensions debate may be a useful focus of attention in terms of weapons development and employment policy [4]. It might well be, for example, that a rule such as 'machines are debarred from making decisions that can be anticipated will result in deliberate or arbitrary attacks upon civilians or their essentials of life' is workable in military ethics, international humanitarian law, and weapons development, production, testing, and legal and regulatory policy. In this connection, a machine can only be said to perform a mission in a morally and legally relevant sense when the machine acts upon instructions and decides that a legitimate target is legitimate, and that its weapons employment is legitimate [5]. In lay terms, this means that it is guided by the moral and legal primers and ethics.

5.2.1 Unmanned Aerial Vehicles (UAVs)

UAV or drone markets are growing like never before. The majority of applications are in the agricultural and construction fields, but some, such as consumer drones, have also attracted considerable attention. Some companies have developed technologies like hydrophobic lens coatings to protect cameras from liquid damage, but most drone-related technologies have added value thanks to measures like image enhancement and altitude restriction during operation. Altitude restriction limits drone flying to prevent airspace hazards, autonomous assault by a third party using the take control function of a drone controller, or unmanned vehicle-based attacks, for example [6]. Research on using a drone swarm alone or with other ground combat robots, including the development of lethal drones, is progressing aggressively across the globe. Timeline considerations aside, UAVs are already providing mobile early warning, reconnaissance, and support for offensive operations in various theaters of conflict. Complementing growth in the consumer drone market, demand is growing for industrial applications among various sectors such as smart factories, manufacturing, and distribution. Measures have been developed to prevent accidents or terrorism using drones. However, like with other technologies, it is always necessary to anticipate not just domestic industrial applications, national demand for monitoring airspace to prevent drone attacks, or concerns about the spread of poisonous substances, but also the possibility of misuse as an autonomous weapon [7].

5.2.2 Autonomous Ground Vehicles

Autonomous ground vehicles are some of the new types of assets in conflict zones. Small robots can carry out missions that can be dangerous and time-consuming, and larger vehicles can be used for teleoperation, targeting, and even firing. Israel has made use of the Harbinger Unmanned Ground Vehicle to detect roadside bombs while keeping troops safe. Similarly, the Foster-Miller TALON system used by various countries and the iRobot PackBot 510 used by the US were deployed at a nuclear disaster to gauge radiation levels and to search for survivors without risking the lives of humans. Automation in warfare must, of course, be balanced with ethical and legal demands and the importance of judgment that humans can bring before firing, but these vehicles have a significant negative effect on humans on missions that are repetitive, dangerous, or mundane [8]. The fears to be addressed about AI proliferation in this scenario relate to the development and deployment of UGVs in locations other than conflict zones. Russia is known to be using UGVs actively to annex territories; their support has been confirmed in Syria, and speculation has not yet occurred behind allegations that Russia has used unmanned ground vehicles to recover a location after a disaster in the UK safely. The British Leaderhouse system combines ground and airborne drones, heavily armed with rifles, records footage, and bombs targets. North Korea and Pakistan have UGV tanks that can shoot automatically, and the Iranian Hezbollah brigades use small UGV boats to patrol the Lebanese coast. Beyond the usage of UGVs in warfare, the broader ambiguous use of these tools exhibits developments at a time of technological acceleration for other kinds of operations [9].

5.2.3 Naval Autonomous Systems

Autonomous naval systems have great potential; first, to support our warships in territory control and defense missions, and second, to extend our warfare approach in critical scenarios. Common autonomous systems in naval operations are USVs, UUVs, and UAVs, which are launched from frigates and deliver real-time sensing, data links, and active defense. In the future, we believe that UUVs and USVs will integrate swarm techniques to accomplish specific assault missions. There are three specific scenarios: replacing anti-submarine helicopters and frigate towed sonar, supporting surveillance, anomaly detection, and physical samples in mining operations, and on-board frigate situation assessment and mine detection. We present some preliminary work on these three autonomous systems [10]. The development of an open, reusable, and maintainable common controlling platform for commercial underwater vehicles integrates intelligent perception, control algorithms, and multi-sensor data processing technologies, and can explore and apply those commercial products in underwater autonomous systems projects that are important because of the great potential of UUVs to improve the performance of our frigates on submarine detection issues. First, on-board UUVs integrate conventional trajectory and GNC methods and deploy on-board Small Kill Vehicles or Narrowband Sonar to conduct real-time contact assessments using multi-information data. Second, UUVs collect multi-dimensional information, perform real-time contact assessments, and anomaly detection using forward-looking imaging sonar with continuous endurance missions [11].

5.3 Technological Advancements in AI

The role of engineers, scientists, technologists, policymakers, and educators is crucial in the design, development, and regulation of AI systems. Further AI-based engineering techniques are likely to be used in civil and military applications in the subsequent years. Before AI development started, the seven base technologies of artificial intelligence were introduced: GPS, the human–computer interface, rule-based systems for activity and fabrication, diagram recognition, speech recognition, and machine learning. The eventual results in the development of AI describe eight grand challenges in AI. These can be tackled with the enhanced resources available today [12].

AI provides insights, tools, and solid information that can ensure more efficient operations and create entirely different defense systems. Using AI, humans can rely on systems that are based on exceptional intelligence, assisting with large-scale unprotected, atypical complications, and improving simulation patterns. It offers new operational concepts and incremental capabilities that re-establish the requirements for tactical, possible, and strategic operations. AI is used to improve the human–machine interface, create security mechanisms, expand sensor networks, secure data, and promote the automation of predictable human tasks [3]. The convergence of AI and education and training has the potential to enhance the Defense Department's capabilities significantly and strengthen the strategic priorities of the United States, including the nation's defense from violence and fear, collaboration, and defense-building to create a free and open environment, and more comprehensive ways to modernize government through sustainable AI systems, as presented in Fig. 5.1.

5.3.1 Machine Learning Applications

As computers grow more powerful, the field of machine learning has boomed. Whatever the future impact of generic AI, greatly refined predictive software has much to offer governments and commercial enterprises today. This section looks at how machine learning is being applied in the defense, security, and intelligence sectors on the world stage, and the national and international initiatives taking place. Defense, security, IT, and intelligence applications require predictive technology that can respond quickly and accurately to the uncertainties of the real world. So it is not

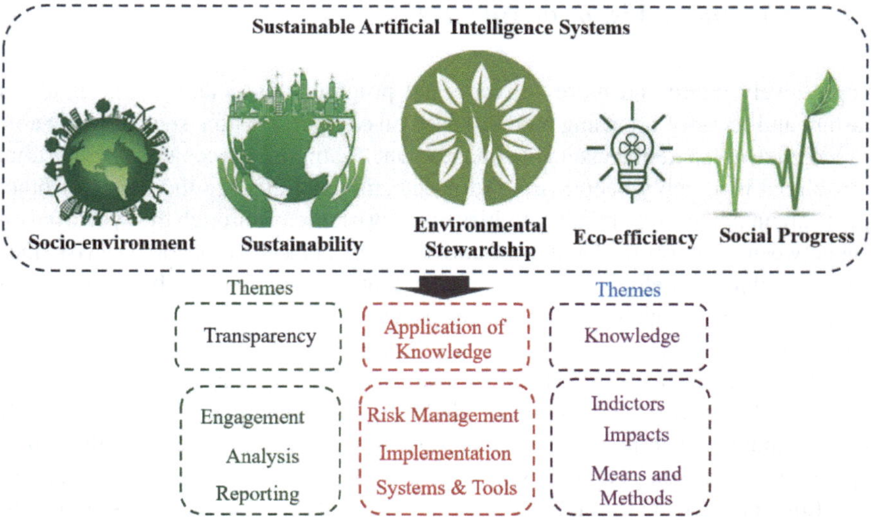

Fig. 5.1 Sustainable artificial intelligence systems

surprising that machine learning meets many more defense and security imperatives than traditional algorithm development. Concrete machine learning applications such as pilotless aircraft and missile systems have shown form for at least a decade, and recently in agriculture [13]. Peer competition between the United States and China, coupled with substantial investment in AI capabilities by both nations, is concentrating AI minds all across the U.S. defense and intelligence world. In mid-2018, the Department of Defense brought together internal staff and private industry to discuss the advanced applications of AI, especially with reference to predictive capability. In the same year, the Office of the Director of National Intelligence held a public summit on artificial intelligence. At first sight, machine learning is about people or machines extracting learned patterns from past data, using those patterns to infer probable responses in new but essentially similar data. The more data is ingested, the more accurate the predictive mappings can become [14]. Enablers of more effective defense and security include data availability and diversity at scale; variable speed of turnover in both data and predictions; implications for human expert trust and reliance; and a level of explainability that, crucially, can extend beyond answers to "why?" These characteristics of machine learning, combined with major investments in the sector by world superpowers such as the United States and China, are giving rise to real-world applications with commercial advantage and enormous competition for assembling this vital talent within the defense and security arenas [12].

5.3.2 Computer Vision in Warfare

Progressively, more and more software and programs, especially in the fields of warfare and security, are using models based on computer vision, such as being able to recognize text and objects inside images. One example is a recognition algorithm whose goal is to apply deep learning for detecting and alerting the police or other agents about potential criminal activities or terrorist threats through the synchronization between its algorithms and the security cameras placed at locations at risk. Due to the fact that the field of view of the paths that security cameras have to cover is relatively large, the quality of the images provided by the cameras is generally low. This means that traditional computer vision models applied to the images coming from the security cameras use features that are insufficient for the identification of people, reading of texts, or recognition of objects within them [11]. Another example is aerial unmanned aerial vehicles, planes, or helicopters equipped with algorithms capable of recognizing anomalies in large crowds, such as public demonstrations, that ultimately provide the authorities with intelligent data that will be used to assist in public safety. Although the growth in the use of AI in warfare and security is relevant and significant, we currently have a shortage of AI programs and, in particular, a lack of a good dataset for this category. The problem lies in the generation of a private and public dataset, in which societal debates are, in many cases, affected by the confidentiality that a treaty imposes, such as those used to determine the limits of the action ranges of the bots in social networks [3].

5.3.3 Robotics and AI Integration

The enormous benefits of using advanced AI in warfare have been recognized by all leading nations involved in robotics and AI research. AI is expected to bring significant changes in the very nature of warfare, which in strikingly different ways will influence political diplomacy, military doctrines, humanitarian activities, and legal and ethical rules regulating the use of force. One can expect that AI, after having been again firing or braking power for military equipment, will be one of the most important battle-support systems envisioned in military doctrines [15]. The development of AI is not only a crucial condition for further integration of military systems, but it is a precondition for creating UROVs, UAVs, UCAVs, autonomous aircraft, amphibious and ground systems, including but not limited to collaborative, cooperative, battlefield robots and RCVs equipped with sophisticated control systems that can autonomously achieve their tasks in a dynamic battlefield environment in the absence of direct supervision by a human operator [5]. Furthermore, these autonomous military vehicles and platforms will not be used only to conduct confined independent operational missions, but as part of battle groups where on-board UGVs equipped with weapon systems and sensors can extend the life of the crew of the bigger manned platforms, such as tanks, ships, and aircraft carriers fighting close combat tasks in remote mode or by using on-board systems with AI. The full integration of systems for mission-critical vehicular networks using full-duplex radio systems will soon require the development of advanced full-duplex communication schemes for vehicle-to-vehicle channels and vehicle-to-BS channels, where a mixture of cooperative communication and other recently proposed transmission schemes, including space division multiple access techniques, are used together for more advanced and sophisticated battlefield applications [6].

5.4 Ethical Considerations

The U.S. Department of Defense released an executive summary of its Defense Science Board Task Force on the development and use of autonomous weapon systems. In the report, the DSB acknowledges broadly that such weapons will change the nature of warfare by reducing the human factor, giving more autonomy to smaller groups of soldiers, or even single soldiers, and offering the possibility of an automatic response to an enemy attack. It also raises the question of differentiating between intent and accidental consequences of these weapons [7]. In Europe, the European Group on Ethics in Science and New Technologies has opined that autonomous robots will be the machines that most challenge the boundary between man and machine, between programmed and non-programmed digging, and therefore the autonomous robots also need attention with a special interdisciplinary approach to make decisions regarding the ethical issues. Unfortunately, the reality as we are witnessing falls way short of addressing the ethical considerations, let alone the profound changes that will inevitably follow. Meanwhile, within pockets of different governments in various regions, there has been an acceleration in research and development of weapons aimed at the capability gap, which aggravates the possibility that lethal autonomous weapons will be produced and placed in theater with ominous consequences [16].

5.4.1 Moral Implications of Autonomous Weapons

Moral and ethical debates about the development of autonomous weapons and the use of AI in warfare have been ongoing for some time. This debate is not unique and actually trails similar debates about automation in other industries. However, the debate over the use of AI in warfare seems especially relevant given the potential consequences of such use. For example, how do we avoid autonomous weapons from making 'immoral' inferences about who or what is a target [17]? How do we ensure that autonomous weapons adhere to international codes of engagement and avoid shooting the enemy when unarmed or when they wish to surrender? How do we ensure that, in the context of the speed of movement and information, proper authorization is made, or how do we implement force control? A further, and even more troubling question to address is whether autonomous weapons can be programmed in such a way to reflect the complexity of international humanitarian law, which is built upon the principles of distinction, military necessity, proportionality, and humanity. Can autonomous weapons ever be a legitimate form of warfare [18]?

5.4.2 *Accountability in Warfare*

According to Article 36 of the Additional Protocol I, "in the study, development, acquisition or adoption of a new weapon, means or method of warfare, a High Contracting Party is under an obligation to determine whether its employment would, in some or all circumstances, be prohibited by this Protocol or by any other rule of international law applicable to the High Contracting Party." Leading the way in the weaponization of AI, Russia and China have both openly expressed their ambitions for military dominance in AI over the United States and beyond. The use of AI weapon technologies in actual battlegrounds might jeopardize civilian lives, security, and public welfare [19]. With the increasing integration of AI in warfare, the accountability of the programmed machines seems to be in need of some formal agreements on both international and national levels, especially when they commit battles, crimes, or atrocities. While such dilemmas still bring significant inadequacies on the responsive liability issue as applied to killing robots, the debate now for some form of compensatory solution or an appropriate division of the reasonable distribution of responsibility may pave the way for the partial adaptation of the existing legal norms concerning the use of violence. The social consequences of AI in the military could have a significant backlash and jeopardize any current fragile agreements [20]. Technological systems, if poorly chosen, can inflict terrible consequences; hence, the focus on safety through collective approaches is presented in Fig. 5.2.

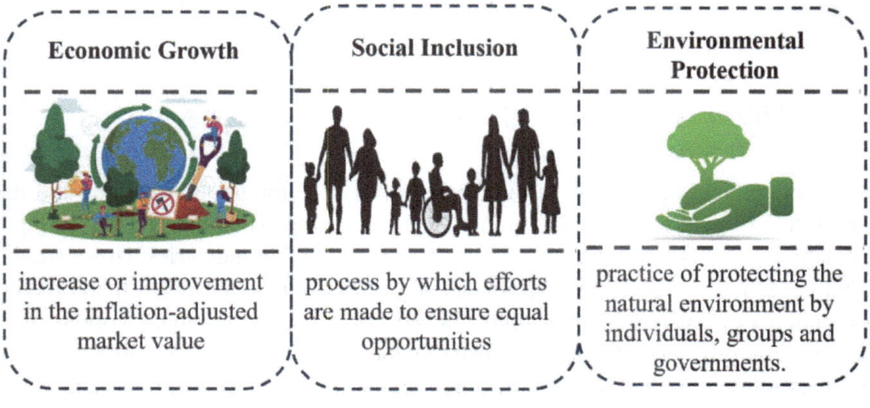

Fig. 5.2 Social artificial intelligence, sustainable approaches

5.5 Legal Frameworks Governing Autonomous Weapons

Reading our recent papers on AI in warfare and security, a follower expressed surprise that we touched so little on the international legal frameworks governing the design, development, testing, deployment, and use of new weapons, specifically autonomous weapons. In light of the fact that legal perspectives remain underappreciated in the ongoing discourse on the ethical, moral, security, and diplomatic challenges and implications of the rise of autonomous weapons and AI, as well as misperceptions of the current state of international law on this controversial issue, we thought it might be a good idea to put together a brief paper for general consumption on the international legal frameworks governing the use of increasingly autonomous weapons, with a view to dispelling contrarian myths on the issue [21]. In the first part of this chapter, we elaborate a little on the status of various international legal frameworks governing the use of increasingly autonomous weapons in the battlefield, including the Martens Clause, the 1983 United Nations General Assembly Resolution, the 1980 U.N. Convention on Prohibitions or Restrictions on the Use of Certain Conventional Weapons Which May Be Deemed to Be Excessively Injurious or to Have Indiscriminate Effects, the 1996 CCW Protocol on blinding laser weapons, and AI-related failed initiatives. In the second part, we discuss the proper role of lawyers in discussions on the ethics and moralities of weaponization and warfare, in light of failed technological utopianism and misperceptions perpetuated by the media [22].

5.5.1 International Humanitarian Law

There is now significant debate over whether autonomous weapon systems can comply with key bodies of international law concerning military hostilities, such as International Humanitarian Law, also often called the Law of Armed Conflict. Many states and legal experts argue that existing international law on the use of force remains relevant, applicable, and sufficient to address the deployment of autonomous weapon systems. They maintain that there is no obligation under international law to ban or introduce specific restrictions on the development or deployment of lethal autonomous weapon systems. With respect to International Humanitarian Law, they have underlined that autonomous weapon systems would have less capability than a human being [23]. It is important to distinguish fully autonomous weapons as distinct from existing platforms that employ automated functions to detect, track, identify, and engage targets without human interaction. International Human Rights Law would form part of the existing legal frameworks. It should be noted that the law of armed conflict must be complied with at all times. Indeed, the concern of many commentators is not the legality of autonomous weapons alone, but rather the techniques of autonomous weapons not being under the effective control of the user. Which state would be held responsible when systems fail to act in line with the laws of war in such a situation? It is a major problem when the unique circumstances of armed conflict could, in some cases, make unmanned system decisions preferable to those of humans. The criteria used to comply with international laws would be rather complex and require extensive data [24]. Given that most participants in the debate consider that compliance with international law is a necessity, the disagreement between the legal and scientific communities concerning the possibility of such compliance is a major challenge that should be addressed.

5.5.2 Regulations and Treaties

International humanitarian law (IHL), the law that regulates armed conflict, is more than a century old. Although often thought of in the context of protecting soldiers, IHL also establishes rules that protect civilians and combatants who are hors de combat (outside the fight, such as the wounded and prisoners). It sets limits on the types of weapons and methods of warfare that may be used, seeking to reduce unnecessary human suffering. The increasing use of AI in weapons has led to a proliferation of international organizations, experts, and others offering an array of opinions on the legal and ethical concerns raised by autonomous weapons. AI in warfare is not just a matter for one nation [25]. Many other nations are already making use of AI for a wide variety of military purposes. There are important global consequences to this greater reliance on AI, including regional instabilities driven by inequality in military capabilities. The impacts of AI in national and international security need to be addressed as a global issue. There is widespread agreement that strong governance is needed to guide the development and deployment of AI for military applications. However, what this governance should look like is heavily debated. Some influential entities in the global technology community have put forth high-level ethical guidelines for responsible AI development that include military AI usage. Expert debates have further delved into approaches such as bans, treaties, verification protocols, and certain prohibitions on use [26]. These discussions must include global perspectives, recognizing the variety of valid viewpoints from non-governmental organizations and those who are otherwise materially affected by the development and use of military AI.

5.6 Global Security Threats Posed by AI

Other global security threats posed by artificial intelligence technologies include the breakdown of global arms control agreements, infiltration and manipulation of social media to incite violence, mass surveillance, and the ability of terrorists to use inexpensive, easily available tools to deploy unmanned weapons. As these technologies cross national borders, they will also allow numerous smaller conflicts to expand rapidly, escalating into regional or global conflicts involving multiple superpowers. Some predict that if AI warfare tools are widely deployed, some national infrastructures could be brought to their knees. Decision-makers should heed these predictions and ensure an understanding of the opportunities and risks associated with AI-related technologies to prepare effective and timely responses to future challenges. In summary, a large number of possible worst-case scenarios need to be analyzed, evaluated, and considered by government leaders, military planners, and philosophers [3]. The above examples serve to illustrate just the most pressing issues confronting us in the coming years. The exploration of fully autonomous weapons and controversial proposals for their use is part of a larger and longer debate about the changing relationship between human beings, machines, and politics. This debate is likely to lead to significant changes in strategic culture and the global balance of power as well. When considering these and other potential powerful capabilities of artificial intelligence systems, it is important to evaluate our heightened security risks in order to determine what appropriate ethical and technological constraints should be imposed for public safety concerns [5]. Small technical glitches that arise from the inherent complexity of the systems could lead to potentially catastrophic outcomes. As a result, institutions responsible for armed conflict management and increased AI oversight and capabilities are crucial.

5.6.1 Cybersecurity Risks

The realm of cybersecurity is currently seeing a significant increase in the number of AI products designed to benefit organizations by augmenting or replacing human actors in the identification and mitigation of cyber attacks. While the development of AI-enhanced technology aimed at the reduction of cybersecurity threats is a promising aspect of the security of networked systems and the Internet of Things, it is also the case that those same technologies can be dual-use in effect; that is, they can be used to threaten, rather than to defend [8]. Some cybersecurity attacks occur in the physical world—attacks on power and water systems, aviation, and military target classification, for example. The physical damage resulting from such attacks has the potential to kill or injure human beings. Meanwhile, increases in autonomous machine decision-making have led to human and computer conflicts in cases such as those involving autonomous cars and robots in warehouses. In short, while increases in autonomy can provide desirable security and safety benefits, they can also come with significant risks, as both humans and machines take on high-stakes security and safety functions [27].

5.6.2 AI in Terrorism

AI has further leveled the battlefield, providing groups seeking to disrupt the international order with new tools and options. Terrorist organizations have long exploited advances in technology to create weapons and tools that threaten more advanced and larger forces. To date, these have been small-scale threats, like the limitations of available resources. There is little doubt that terrorists are exploring the weaponization of chatbots and AI. Cybersecurity analysts hold little hope that they will remain unsuccessful. New weapons like AI-powered tools have the potential to enhance the effectiveness of exposed rail-mounted mine-clearing gear vulnerable to IEDs at a considerable cost of additional weight [18]. As foes develop chatbots that engage soldiers and simulate real communications, they risk redirecting the attention of battle-trained troops, confusing them, and making it difficult for soldiers to detect and apprehend real messaging in a dynamic environment. AI could also allow adversaries to rapidly detect if there is a dialogue with a bot and cover up potential traps by spreading or deleting AI-generated misinformation and fakes. Established algorithms based on carefully pre-established training data and selective data breaches can probably succeed in automatically identifying an adversary developing troops' snapshots. Given that neural networks lack the mechanisms that humans use to rationalize their decisions, troops will have a hard time understanding the implications of dubious decisions made by AI [20]. The demand for moral and tactical responsibility will be further negatively affected.

5.6.3 Geopolitical Tensions

The pandemic has accelerated existing geopolitical tensions. NATO denies defining China as an adversary, and the US administration hopes to restore diplomacy in international relations as well as its role as a democratic leader. The EU administration makes plans about strategic autonomy, emphasizing that the differences between the US and China constitute an obstacle to mutually benefiting from cooperation. The US planned to spend significantly on preparing for competition. G20 members referred to protectionism as a global risk. No one has a clear plan for tapering the dominance of the American dollar and for a compatible solution due to the aftermath of the suspension of trade [21]. Independent of the anti-COVID-19 initiative, the EU is more passive in international politics. The future of the 27-member European Union regarding the rule of law problem, Eastern European, and Western Balkan countries is unclear. Although political talks are encouraged, the Turkish-Chinese negotiations and trade activities strengthen China's regional role in geoeconomic warfare and negotiations bringing the three continents together. The Arab Spring, leading to wars and political fights, has not impacted the continuity of welfare restrictions concentrating on the local tourism route. Exclusive military tourism signifies the global security setting and the problems of local governments in pursuing alternative revenue-generating sources [22].

Securing water interests is of general interest in cases of geopolitical tensions and welfare restrictions among neighbors and further from large empires in connection with global conflicts. The United States uses water as a force multiplier. Opportunities to use hydraulic fracturing technology challenge Russia's economy in Europe, enabling competitive prices in Europe with the Greater Caucasus, the Black Sea, and

Greece. The Arab countries are worried about US defense measures and the Israeli tech industry in security, aerospace, and IoT. Israeli companies generate revenue from exports and collaborations, and they receive security clearances quickly [28]. Israeli cybersecurity expertise is globally acknowledged for agreements mentioning bilateral cooperation, involving critical infrastructure protection. The Joint Committee signed in 2016 involves a Work Group. The Arab official declaration fosters investments from China, and security clearance is difficult to obtain. The consequence is a sensitive relationship, limited in what is published. The GCC countries reacted as they wanted to sign individual agreements with Israel; most of the GCC countries and Israel signed the Abraham Accords [11]. Israel creates trust among allies and achieves agreements during win–win negotiations, decisions, and defense to fight threats such as military and nuclear threats, as presented in Table 5.1.

5.7 Case Studies of AI in Warfare

The arrival of AIs and subsequent information-based revolutions is changing, or has already changed, military practices and has had a transformational and disruptive effect on issues such as strategic decision-making, operational planning, main weapon systems, and support systems. This chapter first discusses how these AIs may be concerned. Then it discusses key characteristics of these AI technologies through the lens of the framework of three attributes: intelligence process, human interaction, and level of autonomy. If the "all" is used, AI technology has the greatest impact on the relationship between human beings and their tasks (including warfare brought about by them) [29]. This new human mission view does not necessarily eliminate battles, as the feedback of these missions to the machines is needed to direct them and win the war effectively. This chapter provides four case studies: the first is keeping a hand on the tip of the weapon, the case of the sword of the crewless submarine; the second is to sit around them, the case of the five alterant anti-fighter aircraft aboard the sixth generation UAV charger; and the case of the relationship between the aircrew and the air operation, the operation of the automatic armed aircraft on board the Ukrainian aircraft. The final case involved the adjustment of the operator in effectively exploiting the robot charging swarming missiles in the entry and control of the air defense at sea, and in the final stage of space development radar. Finally, it introduces the relationship model between the human operator and the autonomous system of fighting joint warfare in future mechanized operations (i.e., ground active mobile platform) [30]. Through these case studies, we summarize the model that defines the relationship between the human operator and the autonomous weapon system of future mechanized operations.

Table 5.1 AI in warfare and security

AI technology	Military application	Potential advantage	Emerging threat	Ethical concern	Global response needed
Autonomous drones	Precision strikes	Minimized human risk	Risk of unauthorized or rogue attacks	Accountability for lethal decisions	Global regulation of drone warfare
AI surveillance systems	Border security and recon	Enhanced monitoring and real-time data	Mass surveillance, privacy violations	Civil liberties infringement	International human rights oversight
Facial recognition AI	Target identification	Faster identification of threats	Misidentification, racial profiling	Discrimination, wrongful targeting	Ethical deployment standards
Swarm robotics	Coordinated attack units	Tactical efficiency and speed	Hard to control, unpredictable behavior	Unpredictable escalation in combat	Protocols for deployment and kill switches
Cybersecurity AI	Threat detection and response	Fast incident response	AI-powered cyberattacks escalation	Global digital warfare	Global treaties on cyber conflict
AI decision support	Strategic Battle planning	Data-driven strategic advantage	Dependence on flawed or biased data	Dehumanization of decision-making	Human-in-the-loop requirement
Lethal autonomous systems	Battlefield combat	Reduced need for troops	No clear rules of engagement	Violation of international laws	UN Conventions on LAWS
Deepfake technology	Psychological warfare	Misinformation for tactical deception	Destabilization of societies	Undermining truth, democracy	Anti-deepfake detection infrastructure
Predictive policing AI	Counter-terrorism	Preemptive threat neutralization	Abuse in authoritarian regimes	Pre-crime ethics, due process	Civil oversight mechanisms
AI-enhanced missiles	Precision navigation	Higher accuracy, reduced collateral damage	Autonomy in destructive power	Moral disengagement from violence	Arms control agreements with AI clauses

5.7.1 Recent Conflicts Involving Autonomous Weapons

The analysis in this research provides specific cases demonstrating advancements made and the deployment during ongoing military conflicts by several nations, including the United States, Israel, the European Union, France, Russia, the United Kingdom, India, Pakistan, South Korea, and China. Other nations with autonomous assets being deployed include Australia, Sweden, Germany, Turkey, Italy, Spain, Iran, Brazil, and Singapore. Before such a high degree of attention on AI technology and the utilization of autonomous weapons, special attention should be paid to these case studies involving semi-autonomous and autonomous robotics during ongoing warfare [27]. The speed and ease with which unmanned systems and AI technology have proliferated have resulted in these technologies being used during recent military conflicts, not only by the superpowers but also by nations with significant economic, scientific, and military potential, assisting in furthering the commercial and military global arms trade. While the intention is not to neglect the role that manned systems play nor to advance any form of an autonomous arms race, the fact remains that semi- and fully unmanned systems are proving themselves and appearing on a large scale. Intelligence, surveillance, and reconnaissance data can supplement and serve as force multipliers if the objective is to conduct global warfare when a nation or group has less-developed conventional and special operations forces [11].

5.7.2 Analysis of AI-Driven Military Operations

The use of AI in military and security issues will define the future of war. It would change character, fundamentally change the dynamics of warfare between states, conflicts within states, and have sweeping implications for international politics and strategies of states. AI would not only revolutionize military operations but also make them more secure, possibly less destructive, and fundamentally different from those of the past. The arms race for AI-fueled autonomous weapons systems has begun, and it is pertinent that the international community needs to consider placing greater emphasis on the advancements in the use of AI and requisite modifications in the existing regulations [31]. The ongoing efforts have the potential to lead to a great transformation in offensive, defensive, and cooperative military strategies. Over the last few years, the military has spent heavily on AI technologies, and they have come a long way. The increased access and open nature of AI technologies enable the development of advanced and capable AI-enabled military concepts. Contemporary and future AI-enabled military applications are generally driven by two end-and-conformance-based perspectives: Reducing or stopping harm, loss of life, and/or ill health resulting from the conflict; and Dispatching with the scale and precision economies of violence—more efficient use of military capabilities. The basic military operations—analysis, planning, deployment, maneuver, support, and engagement—are known to benefit from the AI application, with general improvements in timeliness, quality, and/or rigor in comparison to other instances [23]. However, using AI support for making quicker decisions for AI-automated harm, which would take over full responsibility in the space of support and/or engagement, was a revered border that posed a major global threat. With the rapid improvement of AI capabilities, the simple fact that AI supports a broad range of military operations will soon become relevant. By incorporating AI into existing military applications through collaborative decision-making mechanisms, the low cost of AI intervention and its real-time accessibility encouraged non-state actors to enter the AI-enhanced battlefield. AI-oriented military developments and problems raise several policy questions of national and international interest that are addressed more explicitly [32, 33]. The careful thinking of these questions around military AI use and applications would enable better outcomes.

5.8 Future of Warfare with AI

AI offers new tools not just for achieving better military capability and operational efficiency in traditional domains like air, land, and sea, but also for making progress in important new domains such as cyber. With ongoing breakthroughs in the development of AI, the characteristics of military conflict and warfare are expected to change at a rapid pace. These transformational changes are influenced not only by the capabilities of AI but also by how AI would change the overall geometry of warfare across the landscape of missions. Building on this growing interest and broad range of expectations for effectiveness in concrete applications, most military organizations are presently taking bids to embrace AI research, raise funding for private sector technologies via military and intelligence investment programs, and hire specialists to develop and implement AI for multiple tracks in military operations [6].

The elements of the competition and conflict that AI-enabled tools can help with are numerous. It is the relative weight bearing in military, political, and societal terms that is creating interest in investing in this technology. Discussions about AI in military settings often include opinions about effectiveness in various warfare domains. Factors that can make AI more dynamic in war zones are the speed of advanced computing systems, the multi-mission capacity of mobile but fully equipped soldiers, the capability for AI to learn significantly longer than the typical rotation of pilots, or 'go when wet' for sea vessels. The integration of AI with hardware to maximally use these elements depends on the specific circumstances, but can almost certainly bring important new military capabilities into the modern operation of various naval, aerial, terrestrial, space, and cyber domains [12, 27].

5.8.1 Predictions for Autonomous Weapons Development

Will autonomous weapons develop rapidly enough for international treaties to be effective? AI, robotic, and sensor technologies are advancing rapidly, but predicting the progress and integration of such technologies is challenging due to their complexity. It will take time for the most flexible autonomous weapons to be developed, and such weapons will likely first appear at a major power military comparable scale. How many and how capable autonomy will offer is unclear, let alone the speed and timescales of such developments [17]. Technology is unpredictable; considerable human intelligence and creativity, but also considerable luck and resources, will be required to develop and then keep one step ahead of autonomous weapons. The ubiquity and complexity of dual-use technologies muddle our ability to predict the future. Autonomous weapons application is primarily limited by the research and development investors' willingness to fund and support programs. Focusing on human warfare needs causes and/or the development of such capabilities. A state or other actor with capability, ambition, a perceived need, and a desire to use it, would they respect a treaty if they did not sign or lacked the capacity to monitor it [19]?

5.8.2 Potential for AI in Peacekeeping

The objective of the United Nations is to establish the necessary conditions for the peaceful settlement of disputes among members. In this context, preventing disputes is better than solving them, so AI artillery can monitor international situations by analyzing various information sources and can predict potential conflicts and suggest preventive measures to avoid them. Despite many efforts, peacekeeping activities have also faced many problems, and many lives have been lost or sacrificed. Consequently, if AI artillery is deployed, there will be no human life sacrifices, and the low cost may make it highly efficient. AI artillery can collect, integrate, transmit, and demand information, provide information required by a peacekeeping model, make decisions, and direct forces according to those decisions [24]. As done by peacekeepers, AI artillery can control gunfire, that is, targeting and controlling not only its own firing unit for shooting at close range but also other forces such as the Air Force and artillery forces in conventional locations. In addition to gunfire control and air defense, AI artillery can also be used for many activities such as area familiarization, delivery of force messages, search and rescue assistance, and delivery of situation reports, such as the control of route blocking, the construction of tasks, and camps. The use of AI can help reduce the unexpected consequences or errors of the causes presented. It is also possible to prevent discrimination against certain specific characteristic groups [29]. People involved in peacekeeping are frustrated by the decisions and constraints authorities impose on themselves through arbitrary timelines, numerical goals, or marking a profile on when to leave the crisis area.

5.9 Public Perception and Media Representation

This paper will argue for the need to promote a more public debate that addresses not only the technical and legal aspects, but also the strategic, ethical, humanitarian, and moral dilemmas raised by the advent of autonomous lethal weapons, in order to avoid a technological "arms race" that could lead to serious potential implications for global security. After the fulfillment of certain prophecies and the consequent public alarm, which could be as harmful today as it was towards the middle of the last century, AI shifted to financial dynamics and remained an instrument capable of reassuring a worried public when speaking only of fast internet services or non-formidable weapons. In this regard, most of the scenarios of a dystopian future include the presence of various degrees of artificial intelligence [28]. In various forms of media, a large number of now-denominated public intellectuals have been paying attention and warning about the risks that AI will entail. However, most of the time, their arguments are too technical and chequered to captivate the public. The media, not wanting to go unnoticed, replicates these positions, exaggerates a few themes that are capable of promoting public interest, and finally fundamentally strengthens the sensationalist and frightening discourse, which highlights the current limitations of lethal autonomous weapons, presenting them as elements capable of reproducing an x-rated version, capable of initiating a global conflict or provoking great human disasters [34, 35].

5.9.1 Influence of Media on Public Opinion

Information influences public opinion, and these opinions shape government policies, military, and national strategy. It is in the government's and military's best interest to keep the people informed but also to protect classified information. When information is restricted for any reason, the media in a free state might try harder to obtain it and thus foster adversarial relations with military leaders. In the long run, the military establishment should see that a knowledgeable public is a better disciplined and supportive public. Government and military leaders can influence the creation of public opinion by their established position of power. Journalists have not only the power of the pen and the right of publication in a free society, they also have the power to enlighten, broaden, and maintain public opinion [27]. While military leaders have the power to enlist, feed, arm, and train those who will carry out national policy, journalists have the capability to link these soldiers with the people they guard, protect, and serve. By so doing, the intellectuals of the Fourth Estate also enhance the foundation of a just and free society.

5.9.2 Cultural Depictions of AI in Warfare

Many cultures enacted stories and crafted objects based on their concepts of what the supernatural might be and its place in their lives. These depictions both shaped and were shaped by their perception as well as fears and hopes about the unknown. So it seems logical that a force as potent as AI in warfare would be reflected in the cultural output of humanity throughout its evolution. The earliest marks of a culture are caveman drawings. These usually consist of depictions of the primal activities of these men, such as hunting and mating. This feature of depicting activities is reflected in the other works of the civilization: legends, music, and stories. They document their perception of the world in that particular moment [17]. Tales of demons and angels, for example, reflect on the moral or immoral behaviors in this context. Such awareness of the emerging threats and freedoms in this context can be actual assets in the psychological projection of possible interactions with AI systems.

For instance, the concept of robots, which are human-like artificial autonomous agents, has been around since the ancient Greeks. The writers of antiquity generally gave no hint that they envisioned robots as a source of bountiful benefits or that these creations, which they nearly always regarded as hobbled mockeries of true humanity created by the gods, might hold moral lessons. Their creations of automatons were entertaining theatrical sidelines, as mechanical supporting actors for human puppeteers in morality pageants that came more into focus during the Renaissance. These were presentations in which the robot characters simulating the human condition typically received punishment or divine favor as a moral tale [19]. These were early explorations of the remote idea of robots as figures of terror or with moral lesson potential. Such portrayals reflect on the already established concept, in the collective psyche, that the creation and subsequent rejection of this autonomous entity were crimes against nature.

5.9.3 Collaboration Between Nations on AI Regulations

Countries individually and collectively may contribute to, and may profit from, such bodies. Their experiences in implementing disarmament conventions can provide lessons for establishing an AI-regulating organization. The form could range from a stand-alone body to a broader mandate with an AI regulating subgroup with a seat at an existing organization, and would moderate among the interested parties. This would allow international cooperation in the regulation of AI without creating new international bureaucracy. That being said, either of these alternatives is only likely to be politically feasible if it is also bolstered by a geopolitical order that reflects the shared interests and values of participating countries. Coherence in such institutions comes from the congruence between values and implications of political economies of the critical technologies and the configurations of economies and AI strategies that are currently being shaped. We believe it is important to establish a positive purpose early to emphasize values and norms, not control, and that the values are strong and universal in the desires of partners [24]. The belief that norms are critical to a safe, secure, shared technology is powerful, and it is not hard to find international partners across the rapidly evolving AI technologies. Some question the importance of norms, and at the same time dismantle or undermine the international institutional architecture which will provide for the agreement on these norms in a multipolar world. The partners are very clear about the importance and power of international rules, both sets of open norms between themselves and sets to bind the behavior and pose the less technical, lacking the advantages [36].

5.9.4 The Role of Defense Contractors

The involvement of major defense contractors in policy and decision-making bodies, and their politically powerful place within the government's military and intelligence communities, significantly exacerbates threats presented by these technologies in four ways. First, powerful interest groups suppress critically important debates and discussions on the merits and limitations of these military technologies. Once settled into their roles of execution, such contractors thrive on decisions that expand their roles. In what essentially becomes a self-perpetuating cycle, defense contractors seek to expand either their own functions or government expenditures for national security. These companies are already acting on these functions to increase investment in AI [11]. The growing privatization of military and intelligence functions threatens to create private empires on the power and money realized since the initiation of the War on Terror. Second, to help dictate policy as a matter of national security, seemingly growth-driven corporate advocacy becomes increasingly intertwined with the technological, operational, and financial complexity and importance of vital institutional missions and functions. Despite institutional missions that are publicly understood and constitutionally sanctioned, the Pentagon and the US Intelligence Community rely on defense contractor contributions, experience, and expertise, including R&D and military product sales, for the development of AI in military systems [24]. Exploitation of the rapid advances and potential capabilities of AI as competitive business imperatives establishes contracting companies as leaders in an AI arms race between the world's major powers. These designers and manufacturers have already subcontracted for over 100 projects related to autonomous weapons, asserting to their sponsoring governmental units marked advances in weaponry development and reduction of manpower needs. Defense contractors plan to deploy ground applications for a variety of missions in the next few years [27]. Finally, contractors engaged in advances in technology to build superior tools ripe for jobs.

5.10 Impact on Military Strategy

AI has the potential for significant impact on military strategy and the effects of conflict. Regular and irregular armed forces and intelligence agencies are already deeply committed to the benefits and economies of AI, and as a result, the dependency on it increases over time. A weakness of rapid dependency on advanced technologies and associated literacy and training is that security impacts lag implementation and use. This may be the actual state today. Over and above the number and relevance of deployed platforms, superiority in the management and control of the force gives the greatest advantage. AI, especially in conjunction with quantum computing, brings opportunities to develop force multiplier effects. The major decision, which does not appear to be based on outcome assessment, is delegating decision-making power and authority to machines capable of operating at a much faster time scale than humans. Their pushback is that civilians do not have to be exposed to a machine's decisions that may involve a large number of noncombatant casualties [17]. In the public discussion process, technologists should remember that decisions about the use of force are a moral issue accepted by humans. It may take several generations of societal discussion before AI, in the hands of leaders, would be trusted with that ultimate authority.

5.10.1 Shifts in Tactical Approaches

Technological advancement in artificial intelligence, robotics, sensors, and automation was thought to eventually lead to the development of military systems and technologies that can adhere to the Geneva Convention principles. There is little hope for responsible use of AI in warfare and security in the development and deployment of autonomous weapons. Responsible use of AI in warfare and security in the development and deployment of autonomous weapons is far-fetched. It is argued that conventional warfare no longer involves large-scale deployments as seen in past conflicts. This is evident with significant historical attacks [20]. Tactical use changed and was less expensive. There was a shift from bomber aircraft to cruise missiles, stealth technology, and drones. It is feasible that the future will involve miniaturization of warheads, swarm technology, cyborg mammals and insects, and the arms race focusing on speed and precision weaponry. It is argued that most events involving American forces over the past few decades resulted in a fast-moving and less destructive war. The future tactical use of the military will most likely follow the trend of less destructive forces using automation and AI. These fast-moving forces will not only be able to cross the digital battlespace for the purpose of waging war, but also to control urban crowds [23].

5.10.2 AI and Asymmetrical Warfare

AI as a force multiplier increases the potential of terrorists and other non-state actors. These asymmetric forces, by embracing AI, can offset the conventional power of nation-states. They can promote asymmetric warfare in three important ways. First, it can help them detect and seize opportunities and weaknesses in the military and security posture of the more powerful opponent. Second, it can enable the customization of different platforms and weapons or use particular applications of AI, such as improving the detection of targets, leading to a lower cost of asymmetric warfare. AI can be employed to apply distributed systems and survivable weapon systems to counter the advantages of a more powerful opponent [30]. This challenge forces leading AI developers to collaborate in setting the majority of the technological breakthroughs in the area, in establishing common norms of behavior and respect, and in preventing the spread and illegal use of dangerous AI-based systems and platforms. The central power of nation-states, in collaboration with one another, international organizations, civil liberties groups, corporations, and international criminal organizations, plays a significant role in preventing the emergence of a wide array of security and warfare-related AI capabilities without checks, balances, and containment measures [11]. Because of the transformative potential of AI, it can create an impact on security and warfare, and its properties make it particularly hard to control or regulate.

5.11 Economic Implications of AI in Defense

The deployment of artificial intelligence (AI) in defense is also likely to have significant spillover effects into other broader areas of the economy, ranging from the materials and energy sectors to cloud services and semiconductors. Countries with AI advantages in defense can also leverage this leadership to develop many civilian technologies with cross-domain benefits. At the same time, effective defense AI can be used as a cost-effective technology multiplier, particularly by resource-constrained countries, to help level the global military technology playing field. The performance of a country and its higher-level geopolitical affiliation will increasingly affect the health of its innovation ecosystem, predisposing it toward economy-stimulating dynamic competition or rent-seeking stagnation, with implications beyond merely the national interest [28]. Breakthroughs in military technology can serve as production spillovers and across-the-board demand catalysts, reducing costs and extending reach while stimulating economic growth. In addition, AI in defense innovation can directly and indirectly help the military monitor and secure related critical infrastructure like electrical grids and transportation systems that support the broader economy. Increasing global competition, however, may tilt global technology leadership away from traditional market-focused outputs that support the common good for all residents of the world to defense innovations guided by national interest. A diversified innovation and technology ecosystem underpinning long-term societal benefits will likely be a public good that does not emerge naturally but requires active policy support to keep each nation and economy around the world flourishing and growing [2]. The open question is whether mutually beneficial global rules of engagement can be established that forsake short-term one-sided strategic gains to capture long-term expansion of the public good, through new norms or by fostering international institutions and cooperation that align incentives.

5.11.1 Funding and Investment in AI Technologies

The United States has led the world in developing and funding AI technologies and their recent application to cybersecurity and warfare. In fiscal year 2017, the Department of Defense requested $7.4 billion, a 45 percent increase over the previous year, for uncrewed and autonomous systems. The Department of Defense has also focused on $12.8 billion on AI and related fields such as big data, machine learning, and cloud computing. Many private-sector technology companies have won contracts from the government to provide the U.S. military with AI services and products. Funding for cybersecurity R&D has also increased rapidly, but remains far short of current and projected cybersecurity needs [5]. In 2016, AI start-ups received $4.7 billion in investments—more than three times the amount received in 2012. Many practical initiatives promote the promise of AI for developing countries. For instance, a global cybersecurity capacity center has been tracking the rapid growth of countries using AI to improve cybersecurity. AI products use advanced algorithms to help power systems contribute to a low-carbon economy in Tunisia and other developing countries. Small and medium-sized enterprises solve cybersecurity challenges in the Latin America and Caribbean finance industry [7]. These types of initiatives deserve continued support, but to this point, they remain relatively rare.

5.11.2 Cost–Benefit Analysis of Autonomous Weapons

Before autonomous weapons can be placed on the battlefield, the associated risks and benefits should be carefully considered and evaluated. The following paragraphs aid in undertaking such an assessment. The major advantage of autonomous weapons is the prospect of a dramatic increase in the speed of action and the precision of targeting. There are, however, limitations to the extent of these advantages, given that autonomous weapons do not avoid damage to objects whose nature has not been previously fixed; their capabilities are based not on the destruction of a given object, but rather on recognizing the object and the conditions and parameters of the attack. Numerous plans exist to create autonomous weapons, but each of these plans requires a thorough cost–benefit analysis [27]. Such a consideration is distinct from the consideration of the technical, tactical, strategic, and economic advantages and disadvantages of various present and future weapon systems, but it is essential in responding to legal and ethical concerns. First and foremost, the cost of an autonomous system includes assessment of its reliability and predictability, which characterizes its effectiveness; it should be greater than that of comparable weapons. The presence of any predictability and reliability problems in the robot's functioning implies stopping its use; systems without such problems will have substantial job capabilities that can lead to an overall decrease in the investment required for the increase of military effectiveness [3, 9].

5.12 Conclusion

Will the deployment of AI in military and security domains result in Armageddon on Earth? The answer is negative. Since humans are involved in programming and deploying combat AI, reasonable human choices can and will be made to avoid the adoption of overly risky AI decisions or to draft international agreements that would prevent states from such dangerous deployments. The application of commercial AI to military and related tasks should not be faster than the possibility for civil society to adjust to such employment and to develop accountability and responsibility, whether for company decisions, governmental decisions, or those undertaken by professional soldiers and private military companies present in conflict theaters. Criminal laws and obligations for the protection of private property must be analyzed, and the defense of digital property and AI-operated weapons must be incorporated in security conversations and regulations. The inclusion of all stakeholders in such debates is crucial. In this challenging and interdisciplinary environment, more research concerning the use of AI-operated weapons should seek to identify the expected devastation caused by gunfire and to establish the rapidity of the attacks to which such weapons would be more likely to lead. Social, ethical, legal, and psychological evaluations are necessary regarding the use of AI-assisted technologies in defense and military environments, as well as the possible creating of men without sins used in future military theaters. After the technical aspects of deployment have been summarized, we describe in the following paragraphs some recommendations and potential provisions to be included in AI-related treaties.

References

1. Shafik W. Artificial intelligence and machine learning with cyber ethics for the future world. In: Future communication systems using artificial intelligence, internet of things and data science [Internet]. Boca Raton: CRC Press;2024. p. 110–30. https://doi.org/10.1201/9781032648309-9
2. Moustafa N. A new distributed architecture for evaluating AI-based security systems at the edge: Network TON_IoT datasets. Sustain Cities Soc. 2021;72.
3. Elliott D, Soifer E. AI technologies, privacy, and security. Front Artif Intell. 2022;5.

4. Shafik W, Matinkhah SM, Ghasemzadeh M. Theoretical understanding of deep learning in UAV biomedical engineering technologies analysis. SN Comput Sci. 2020;1(6).
5. Pieters W. Explanation and trust: What to tell the user in security and AI? Ethics Inf Technol. 2011;13(1).
6. Alhajeri AAAMA, Safian EEM. Factors of the use of AI technology influencing community security in UAE. Int J Sustain Constr Eng Technol. 2023;14(3).
7. Park YJ, Jones-Jang SM. Surveillance, security, and AI as technological acceptance. AI Soc. 2023;38(6).
8. Xu Y, Bai T, Yu W, Chang S, Atkinson PM, Ghamisi P. AI security for geoscience and remote sensing: challenges and future trends. IEEE Geosci Remote Sens Mag. 2023;11(2).
9. Kalodanis K, Rizomiliotis P, Anagnostopoulos D. European Artificial Intelligence act: an AI security approach. Inform Comp Sec. 2024;32(3).
10. Shafik W. Cyber security perspectives in public spaces: drone case study. In: Handbook of research on cybersecurity risk in contemporary business systems. 2023.
11. Thornton R, Miron M. Towards the 'Third Revolution in Military Affairs': the Russian military's use of AI-enabled cyber warfare. RUSI J. 2020;165(3).
12. Chen T, Liu J, Xiang Y, Niu W, Tong E, Han Z. Adversarial attack and defense in reinforcement learning-from AI security view. Cybersecurity. 2019;2(1).
13. Shafik W. Biosensor-based drones anomaly detection integration for sustainable agriculture development. In: Achieving food security through sustainable agriculture [Internet]. IGI Global;2024. p. 283–316. https://doi.org/10.4018/979-8-3693-4240-4.ch012
14. Shafik W. Predicting future cybercrime trends in the metaverse Era. In: Forecasting cyber crimes in the age of the metaverse [Internet]. IGI Global;2023. p. 78–113. https://doi.org/10.4018/979-8-3693-0220-0.ch005
15. Mohammadi M, Sohn I. AI based energy harvesting security methods: a survey, vol. 9. ICT Express;2023.
16. Frank AB. Gaming AI without AI. J Defense Model Simul. 2022.
17. Arya S, Sharma G. Generative AI images and indian media industry: an overview of opportunities and challenges. J Commun Manag. 2023;2(04).
18. Welch JP. Drone warfare in transnational armed conflict and counterterrorism. J Intell Confl Warfare. 2021;3(3).
19. Knight W. Russia's killer drone in Ukraine raises fears about AI in warfare. Wired. 2022.
20. Nalin LCA, Tripodi P. Future warfare and responsibility management in the AI-based military decision-making process. J Adv Milit Stud. 2023;14(1).
21. McKelvey F, Packer J, Reeves J. AI and the automation of warfare Can J Commun. 2022;47.
22. Konigsburg JA. Modern warfare, spiritual health, and the role of artificial intelligence. Religions (Basel). 2022;13(4).
23. Lewis L, Ilachinski A. Leveraging AI to mitigate civilian harm. Center for Naval Analysis (CNA). 2022.
24. Lu M, Qiu JL. Empowerment or warfare? dark skin, AI camera, and Transsion's patent narratives. Inf Commun Soc. 2022;25(6).
25. Shafik W. Ethical and legal considerations in artificial intelligence. AI-Enab Threat Intell Cyber Risk Assess. 2025;90.
26. Shafik W. Toward a more ethical future of artificial intelligence and data science. In: The ethical frontier of AI and data analysis [Internet]. IGI Global;2024. p. 362–88. https://doi.org/10.4018/979-8-3693-2964-1.ch022
27. Wang W, Zhou H, Li M, Yan J. An autonomous deployment mechanism for AI security services. IEEE Access. 2024;12.
28. Rossiter A. AI-enabled remote warfare: sustaining the western Warfare paradigm? Int Polit. 2023;60(4).
29. Scott K. Reith, Russell, and the robots: AI, warfare, and shaping the debate. In: European conference on information warfare and security, ECCWS. 2022.
30. Allen GC. Understanding China's AI strategy: clues to Chinese strategic thinking on artificial intelligence and national security. Center for a New American Security;2019

31. Yankoski M, Scheirer W, Weninger T. Meme warfare: AI countermeasures to disinformation should focus on popular, not perfect, fakes. Bull Atomic Sci. 2021;77(3).
32. Shafik W. Deep learning impacts in the field of artificial intelligence. In: Deep learning concepts in operations research [Internet]. New York: Auerbach Publications;2024. p. 9–26. https://doi.org/10.1201/9781003433309-2
33. Shafik W. Generative AI for social good and sustainable development. In: Generative AI: current trends and applications. Springer;2024. p. 185–217.
34. Shafik W. Community and Artificial Intelligence-enabled disaster management and preparedness. In: Navigating natural hazards in mountainous topographies: exploring the challenges and opportunities of living [Internet]. Springer;2024. p. 243–66. https://doi.org/10.1007/978-3-031-65862-4_13
35. Shafik W. Artificial intelligence models to prevent forest fires. In: AI and IoT for proactive disaster management [Internet]. IGI Global;2024. p. 78–106. https://doi.org/10.4018/979-8-3693-3896-4.ch005
36. Hageback N, Hedblom D. AI for digital warfare. 2021.

Chapter 6
Misinformation and Manipulation: The Dark Side of AI in Media and Politics

6.1 Introduction

Misinformation is defined here as false information that is spread, regardless of intent. Misinformation can originate from a variety of sources, including false rumors, errors, satire, and hoaxes. Unintentional misinformation can originate from any person or group motivated by an incentive, or it can spring from simple human error. For this reason, the countermeasures against misinformation often differ from those of disinformation. Disinformation refers specifically to false information that is spread with the intention to deceive or manipulate [1]. Its intent can be to influence public opinion, financial markets, or consumer behavior, or to manipulate public policies and societal conflict. Many online information manipulators are motivated by financial incentives, such as increasing web traffic and advertising profits across thousands of disguised social media accounts. Disinformation actors also include nation-states and other institutional actors working to gain a strategic advantage by manipulating public opinion or causing confusion around specific political, economic, security, and social issues. They can mobilize bot armies and troll armies, create fake social media profiles, disseminate manipulated images and videos, and fund community organizations that masquerade as being from a particular country when they are not [2].

6.1.1 Definition and Types of Misinformation

Misinformation in Media and Politics: Misinformation, and the closely related problems of disinformation and malinformation, have become more evident in the media and politics in recent years. Some scholars have even called it an "infodemic." As we rely more and more on technology for our information, trust in that information is eroded. Now, it is not just the spread of misinformation that is called into question

W. Shafik, *The Dark Side of AI*, https://doi.org/10.1007/978-3-032-09130-7_6

but also the mechanisms that enable that spread. The advent of artificial intelligence has enabled new systems to manipulate vast amounts of information rapidly and at relatively low cost [3]. Misinformation is, very simply, false or inaccurate information that is spread regardless of intent to deceive. Disinformation goes a step further and is false information spread with the intent to deceive. Finally, misinformation is the distribution of private information regardless of the truth, whether the intent is to embarrass or harm someone. There is a wide array of ways to generate or be exposed to misinformation. It can be by the entire text, or by a word, or just a video. It can involve identifying manipulated photos or videos and recognizing messages spread to detract from election integrity and to manipulate public opinion about public policies [4].

6.1.2 Historical Context and Evolution

As attention to AI misinformation and disinformation has grown, so too has an understanding of its role. The concept of today's so-called deepfakes is anything but new. In fact, the widespread use of misinformation and disinformation dates to at least the fifteenth century, when the invention of the printing press made it possible to mass-produce works of infotainment, easily facilitating the spread of both entertainment and contemporary citizen journalism. The twentieth century revealed how the mass media could be manipulated to serve the needs of autocracy and propaganda. And in the twenty-first century, the internet has become an essential tool for disseminating biased or false information, often created without the need for journalistic responsibility. While social networks have imposed algorithms or detection mechanisms in an attempt to curb broadcasted misinformation, the use of deep learning algorithms to imbue AI with the potential for breaching security in combination with convincing audiovisual forgery, known as deepfakes, has added a strong dimension to the misinformation phenomenon [5].

The concept of deepfakes has manifested specifically in content verification, manipulation, and protection, emanating from the use of AI as a source of either good or harm. Indeed, the inception of the term was based on the emergence of realistic deep learning falsification methods in this field. As a social network user, it is difficult to distinguish between a doctored deepfake created using imitation learning of a person's speech or identity and a deepfake created from a set of visual, audio, or written data [6]. In contrast to a cheap fake, a deepfake can be generated quickly and without in-depth machine learning knowledge, enabling straightforward content or message falsification. Given that the number of unique visitors to deepfake websites grew significantly, and perhaps the reports of malicious and non-consensual use of deepfake algorithms against women in some parts of the world, it is apparent that society as a whole should assume that users are being exposed to a potentially harmful stream with a high velocity, even if the likelihood of being deceived is low [7].

6.2 The Role of AI in Media

In the age of global digitalization, the news media have been influenced in various ways by the digital transformation. The production and distribution of news content have changed fundamentally towards transmedial and multimodal content types, increasingly generated by AI-based tools. Furthermore, the nature of audience interaction with media content in this environment has become much more complex because of the growing problem of information pollution. The shift in the business model in favor of user-published content, the employment of propaganda bots in social networks, and the efforts of misinformation actors in influencing search engines and news aggregators have all changed the ecosystem of online media in favor of political manipulation at the expense of journalism, as news platforms have stopped being trustworthy gatekeepers of reliable news [8, 9]. Media have their responsibilities as communicators, interpreters, and watchdogs. The transformative power of AI has brought new power into the hands of media organizations, which also have significant societal responsibilities. AI has captured the attention of the tech media, with advances at the skill level being reported from all areas of the media, from data preparation to the effective delivery of the news. Algorithms have become quicker at crunching data, with personalization leading to increased reader engagement. The rise of smart assistance has created a new opportunity for interaction with readers, reflected in the development of demand-aware gradients, a new research trend called end-to-end personalization. But no space is neutral, and the algorithmically managed media space can easily become a brutal one. Today's media space has given a voice to anyone with internet access [10]. The result is a new and unpredictable landscape, one where interactions can produce both citizens and non-citizens, and where the so-called end of gatekeepers has given way to the curious transformation of gatekeepers into gatekeepers, using ChatGPT as presented in Fig. 6.1.

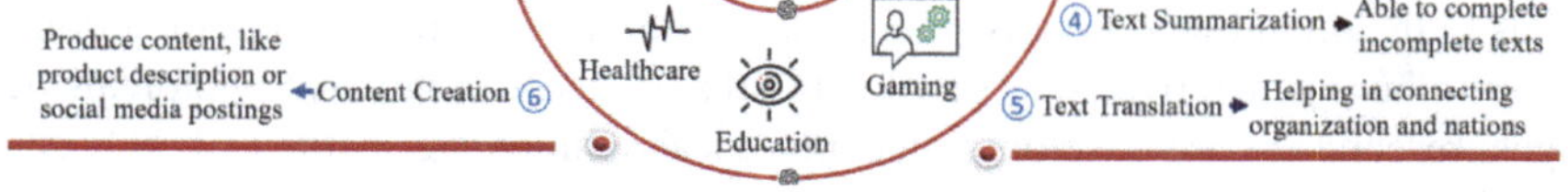

Fig. 6.1 ChatGPT application

6.2.1 AI Algorithms in News Distribution

The problems related to disinformation and propaganda can clearly be seen in the field of news distribution. What makes this technological revolution of the news distribution system so relevant to us, and why is this topic a novelty in this respect? The answer ensures the whole heart of the research, which argues that constant technological transformations and innovation outbreaks in the news distribution system compromise the reliable information that citizens have of collective affairs and that the political agenda is frequently influenced and controlled by the interests of information distributors, especially in the case of large digital corporations [3]. In fact, it is actually the algorithms that some mass use in social networks, allowing a greater reach, reducing the barriers to publication, increasing people's audience, and allowing indie producers to compete symmetrically with other types of producers. The questions are the level of control of manipulatory practices that these tools can allow to be performed, and the existence of agents and institutions within the industry that can reverse these situations. For this, research is conducted in a journalistic news producer, relying on means of production and attention management as objects of analysis [11].

6.2.2 Content Creation and Deepfakes

The involvement of AI in content creation has been increasing over the past decade. Both news organizations and the general public are benefiting from the speed and reliability of AI-generated content, especially in breaking news situations. However, the power of AI in image generation has a downside as well. Deepfakes can be a significant threat to society by generating realistic-looking but false images and videos [12]. While the propaganda effect of such creations has been known for a long time, the democratization of the technology, making deepfakes available to the general public or actors from the entertainment industry, will have repercussions that can hardly be foreseen. This development, hitting the realm of politics, just in time for the upcoming presidential elections, is both unprecedented and alarming [13]. Deepfakes have been coined the weapon of mass mayhem and can indeed become a severe security threat once utilized to spread disinformation. As with most innovations in the digital world, tools such as deepfakes are effective in both directions. Similar to how generative adversarial networks are used to produce photorealistic images, researchers are developing technological solutions to distinguish between original and manipulated content. In education and journalism, such detectors can be of great benefit in contrasting false content, such as counterfeit papers, instructions, or comments on the educational path, or humanoid AI news anchors [14]. Furthermore, one must not forget the potential for positive uses of deepfakes in areas including education, entertainment, history, and politics.

6.3 Political Manipulation Through AI

The previous section made it clear that political deception in general, and election manipulation in particular, have both a rich history and are a serious modern challenge in Western democracies. There are big and important questions here, including whether the old categories of 'disinformation' versus 'propaganda' versus 'spin' are sufficient for a fast-moving, polarized age in which humans see algorithm-driven content. Are we even capable, as yet, of distinguishing between 'spin' on the one hand, and purposeful 'lying' on the other? More fundamentally, as critics pointed out with regard to the admittedly sharp definition of 'Fake News,' posing this as a question of taking down content overlooks the potential problems associated with digital platforms, which prioritize, contextualize, and promote information through an algorithmic 'black box' [15]. This chapter is devoted to unlocking that black box by considering in some detail the dual role of artificial intelligence in amplifying political messages both in media and the political sphere. It does not take a conspiracy theory to explain why many representative governments take their media so seriously. Apart from the obvious fact that the media is the main interface between leaders and the led, a lesser-known fact is that political stakeholders are primary consumers of news; ministers come first, and the rest of us on the coattails. Moreover, the news is an immensely powerful 'political influence ecology'; it helps shape the political understanding of voters, encourages trying governments, and it is fluid to respond when news makers think they might do better by electing different representatives [16].

6.3.1 Campaign Strategies and Targeting

Social media, AI, and data science have transformed political campaign strategies. The algorithm and other advanced approaches are used to identify "influentials" who maximize the spread of messages in cyberbalkanized networks and "influenceables" who are susceptible to persuasion. The latter group can be tallied on one's available followers, fans, and lists, segmented by country, locality, demographic, topic interest, sentiment, and other microtarget attributes. This enables campaigns to enhance their engaging content, "fake news," disinformation, trolling, and other harmful tactics that have now been professionalized to the level of other industries [17]. Unfortunately, unlike medical ethics in AI, there is currently no political code of professional responsibility to protect the weak and harmed, effectively increasing the potential for the exploitation of followers due to essentially being unwatched by social media platforms and other regulatory authorities. These techniques are intended to suppress and manipulate the independent voice and vote, driving people apart and making them more passive, nonvocal, noninformed, mentally dependent, and emotionally controllable. The combined use of AI and behaviorism to continuously monitor and adjust persuasive microtactics in the form of manipulation, persuasion, and obedience is undermining liberal democracy and suffocating the very personal skills that society needs to develop strong, independent, open-minded, societal, and trustworthy human beings [18].

6.3.2 Disinformation Campaigns

Disinformation campaigns aim to manipulate opinion, sow dissension, and confuse in the context of existing social, political, and military struggles. The potential of AI makes them far more dangerous. AI empowers disinformation campaigns to attack at a never-before-seen scale, posing an existential threat to functioning democracies as we know them. AI models create crafted content to deceive, polarize, and manipulate with unprecedented precision. In the recent past, traditional countermeasures aimed to reduce vulnerabilities, promote good information practices, and increase information literacy, although there is a growing acknowledgment of the attributes of the information environment [19]. The information environment is an integral part of a complex environmental system. It consists of the communities of senders, their messages, the receivers, and all the other entities that may intervene to influence message and receiver behavior in ways that may cause consequences that are different from the ones that would have occurred in the absence of intervention. Sampled AI tools are presented in Fig. 6.2.

The common result characterizing manipulation is a distorted perception by the receiver of the actual state of facts. This distortion can lead the receiver to change their actions. AI and cybersecurity researchers are working on technical solutions that would assist in identifying synthetic and human-generated text. Few are investigating processes and practices that integrate non-governmental actors and governments to create opportunities for the people to navigate the information environment more effectively. Governments have started to implement measures that add a layer to an otherwise industry-dominated field. Such measures include online transparency in political advertising and online integrity [16]. Comprehensive legislative frameworks provide opportunities for non-governmental actors and governments to collaborate on regulation, bringing the wisdom of the crowds and best government practices together.

6.4 Social Media and Misinformation Spread

Social media platforms, like Facebook and Twitter (formerly X), have become primary vehicles for the spread of news-related information, especially for large portions of younger generations. Not incidentally, while the overall volume of information spread on those platforms has exploded, so has the phenomenon of misinformation, with side effects on societal discussions and decision-making. Several studies and reports have shown that users on social media can be easily exposed

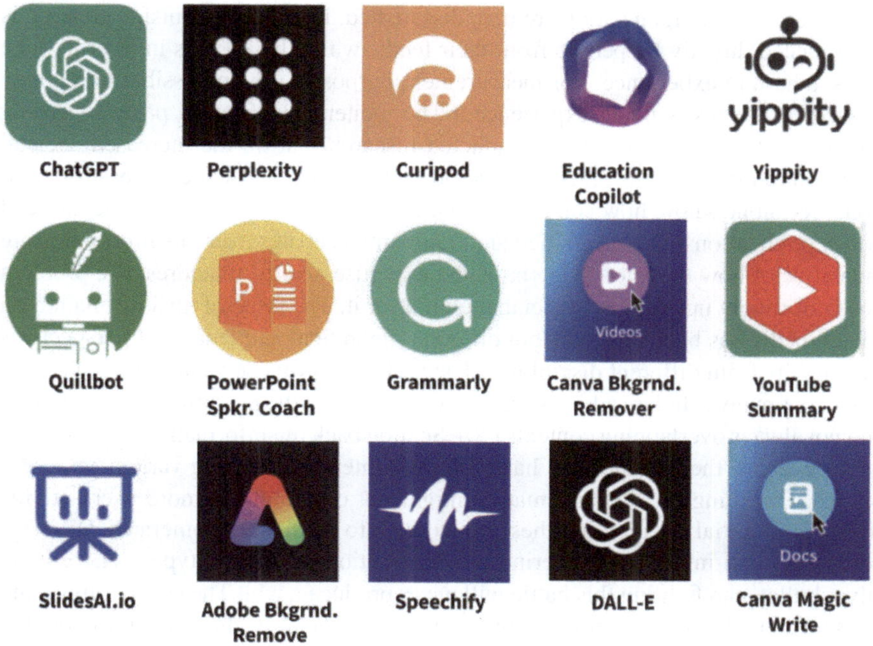

Fig. 6.2 AI tools

to false narratives and that this can have a strong influence on their beliefs. This is especially true when groups of users are not exposed to information that contradicts or corrects the misinformation they are exposed to, or when that information appears as a first contact with the theme [15]. Social media are becoming a privileged space to affect epistemic authority and the social value of some pieces of knowledge. As of 2019, more people reported getting their news from social media than from print newspapers. According to a study on the state of news media reporting, 20% of Americans said that social media is a very important or an important way to get news. At the same time, a public news organization has lost roughly 25% of its listener base in a decade, decreasing from 20% in 2008 to 14% in 2018. Consequently, social media and technology companies are realizing that they are becoming news companies and that they must do something to solve the growing crisis of misinformation online [11].

6.4.1 Platforms and Their Algorithms

In response to growing public dismay over fake news, many social media platforms have increased their efforts to detect and filter it. Although these efforts have resulted in less misinformation being spread, many users, researchers, and nongovernmental organizations remain dissatisfied. Each case of misinformation is experienced directly by people from their feeds, while the success in filtering fake news is hard to experience. For measurement purposes, it is impossible to see how missed fake news is being experienced. The incitement of moral panic regarding misinformation is not based on conducted research, but on the increased societal attention to the problem [2]. Before the problem is properly scoped, ideas for solutions, let alone know-how for these panaceas, are dispersed in policy debate and policy publications. The diversified and fragmented debates over misinformation are indicative of how the knowledge base and expertise needed to address the problem have not caught up with the societal attention for it. The issue of misinformation on social media has been studied from different theoretical and empirical perspectives by scholars from different disciplines. The variety of proposed solutions is a consequence of many scholars addressing only one aspect of the phenomenon in different and not always overlapping contexts [20]. Scaling back misinformation is not simple, partly because these companies have different interests in taking various responsibilities. Providing quality information might not contribute to more user activity, which is essential for growing these networks into media conglomerates. Different companies can invest in countering misinformation of various types. There is no silver bullet, and fighting this battle will never produce a win. The sheer volume and the speed at which information is disseminated on social media make it impossible to neutralize every fake news story with an algorithm [13].

6.4.2 User Behavior and Engagement

The AI and algorithmization of social media platforms require further user engagement with how these technologies work. The recommendation and ranking algorithms make way for audience behavior to engage, react, click, share, or comment on others' posts, and thus contribute to the production and spreading of content that would further trigger these algorithms' engines. However, the right to the future tense is in jeopardy: on one hand, end-users' metrics, log-in actions, and so on would be turned into input data for AI to calculate and predict users' future engagement and choice model; on the other hand, social media users' engagement in this AI era brings more negative dimensions in terms of consumption pathology, filter bubbles, and echo chambers, polarization, and so on [15]. The long-debated over-personalization and data-driven customization in the AI and algorithm era can arguably influence users' information or entertainment seeking behavior through filtering, curating, and prioritizing updates, posts, and even influence the way users see what is going on in the world. The hyper-engagement on these created or established filter bubbles and echo chambers has caused a collective political mental state, the reinforcement of stereotypes and misguided perceptions, and even further political disintegration and polarization, as social media's information-seeking behavior has presented symptoms more akin to an addiction similar to other disruptors such as smoking and obesity. Social media's prioritizing score metrics have turned into the quantification criteria and measurement tools for users [16]. The more online attention and engagement you give to these content producers, the more prioritization for their political stridency, hyperbolic intensity, emotional resonance, or content based on one's bias or prejudice will have a higher chance to be presented within the personal newsfeed and be taken into account by these algorithms [21].

6.5 Case Studies of Misinformation

This power was on full display a day after the November 2019 UK general election when a morning show saw its highest rating of the year as a video clip and AI-subtitled version of Labour Leader Jeremy Corbyn's speech was widely circulated, falsely showing him as a sore loser admitting fault for the party's defeat. This is part of a larger campaign that is as much the creation of the media and elite discourse that criticizes the party as being repelled by its electorate. This campaign even fooled half a dozen of the country's broadsheet editors during the early-morning newspaper conference on election day, and a political figure explains why each platform has a specific function that is used to communicate different messages around any given issue [17]. Similarly, the pejorative encapsulation that surrounds the "dangerous" Corbyn and "toxic" Labour becomes highly weaponized when considering how they are used, and by whom, across social and traditional media in order to construct different understandings of the key issues that, in turn, shape the democratic attitudes of millions of people in the UK. At the same time, compared to the ridiculous speed with which the authorities were able to construct the counter-narrative around the interference in this election, it makes it impossible not to conclude that those same powerful elites are already pressuring the government to intervene across social media and limit potential damage that this report may have, revealing the key mechanism of elite safeguarding; mostly the right-wing press's latest obfuscation and non-disclosure of wrongdoings that the UK public is forced to tolerate [22, 23].

6.5.1 2016 U.S. Presidential Election

There is growing evidence showing how the spread of misinformation has been used in attempts to manipulate public opinion in various countries. One of the most visible and shocking incidents of this was during the 2016 United States Presidential Election, in which many different manipulative tactics were employed through widespread misinformation. Online, fake news articles regarding the fabricated health of Hillary Clinton were some of the most shared news, both on social media and other online platforms. Despite a lack of credible sources, poor language, and outright falsehoods, these stories were collectively shared hundreds of thousands of times. Two media tools backed by the Russian government were popular online sources

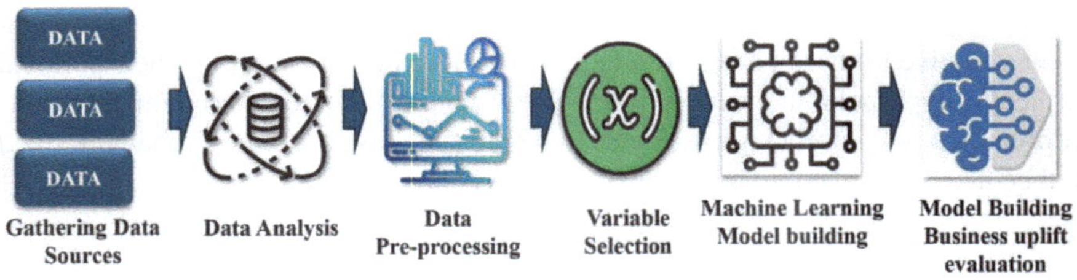

Fig. 6.3 AI model development processes

[20]. Both Russian President Putin and Clinton confirmed Russian interest in interfering in the U.S. election, yet the real effect of such interference and misinformation was not verified. Analysis from a cybersecurity firm found that automated accounts, together with inauthentic websites amplifying certain stories, spread anti-Clinton misinformation during an extremely short burst of activity [24]. The sniper story also saw substantial coverage by foreign manipulators. Following an analysis identifying a lead in tweeting links to sniper and other controversies early in October 2016, the research resulted in discussions about collaboration and was quoted in a story regarding propaganda in a potentially related context [25]. In contrast, most of the other stories were more sophisticated and were often later amplified by content-focused websites or forums with stronger credibility and higher visibility in the U.S., suggesting a more organically successful deployment, as development stages shown in Fig. 6.3.

6.5.2 COVID-19 Misinformation

One of the biggest impacts of COVID-19 is not only the collapse of healthcare systems and economies all around the world but also the universal dissemination, uncontrolled spread, and wide belief in misinformation bordering on conspiracy theories and intentional disinformation based on ignorance and malintent. A significant lens for the analysis of the alterations in public opinion and mass social behaviors is social networks, where traditional mass media and social media interact and influence society by highlighting social problems, making some problems visible, and solving others in order to keep the system of human society sustainable [26]. With the prevalence of AI as a media moderation tool, it is strategically important to retrieve knowledge on the gap that exists between the potential of deep learning methods and the results achieved in real democracies. AI systems are largely responsible for the creation, distribution, and amplification of fake information coming from the manipulation of images and text, and play a significant role in allowing the spread of disinformation about COVID-19, which results in shaping public opinion and hence society [2].

The use of pre-trained Convolutional Neural Networks (CNN) and Recurrent Neural Networks (RNN) is coherent. It is demonstrated that pre-trained state-of-the-art CNN models are capable of effectively classifying fake health content by using both the text and images conveyed in the conversation tweets, showing the ability of AI to be a potentially valuable tool in the fight against COVID-19 misinformation. Furthermore, it is arguably demonstrated that state-of-the-art pre-trained RNN models are able to effectively classify the stance of tweets about health-related pieces of news when used sequentially from different pre-trained models, exposing a significant issue in both public opinion and the regulation of social media [27].

6.5.3 Global Perspectives on Misinformation

There are a few specific cultural, political, and social components that drive people in different countries and regions to seek out information from certain sources, share that information within their social networks in ways that promote engagement and group affiliation, and believe what they are reading. These components are what drive the emotional nature of the so-called 'filter bubbles' and 'echo chambers'. Similarly, inherent in this persuasive approach is understanding the motivations for those outside of the country of origin who wish to effect change in a specific country. In this chapter, we learn about these factors present in the ongoing Brexit debate by studying the online communities present in the UK and a small number of other European countries [28]. The literature strongly suggests culturally and geographically integrated people will interact more if they have similar knowledge and are socially related. This chapter investigates how different individuals from specific countries are motivated to act in the global public sphere. They are united by political and social interests related to Brexit and motivated in doing so by psychological, social, or political components. These include an affiliation, at a high level, with the idea that the European Union is universally evil and a preference, at a low level, for the impact of the opinion created [29]. We also find that individuals are more interested in locally relevant Brexit referendum content and migrate directly to their trusted politically aligned news brand away from neutral web search to seek out their globally reminded fact within the specific country's national media organizations.

6.6 Ethical Implications of AI in Media

These risks and the potential deleterious impacts on democracy make it all the more important that AI in media (and in politics) be subject to the sorts of ethical considerations, public debate, and regulation that have generally been absent from AI investments to date. While optimists of AI technologies might argue that by understanding their architecture better through regulatory means or through internal design practices, the benefits of AI technologies have the potential to be unlocked more easily, the power of these technologies for misuse does give us pause. With the rapid rate of AI advancement and its already clear deployment in media, the time to talk about how to protect the health of our public sphere and the minds of our citizens is upon us [30]. The purpose of this chapter is to raise awareness around the negative implications of AI use pertaining to these fields and to caution scholars and fellow academics to work diligently to ensure that the promise of AI to democratize information and increase knowledge is not co-opted by these negative implications. It is believed that understanding is the first and best defense. The essay proceeds in the following way. First, the basic technologies that make up media and AI systems are identified and analyzed. This is followed by a discussion around specific uses of AI in media that are harmful to individuals and society [31]. In the fourth section, conclusions that arise from the earlier analysis.

6.6.1 Responsibility of Tech Companies

Most importantly, tech companies need to treat misinformation and manipulation as political and propaganda content, not as quality control or as content that needs to be regulated in the way that commercial advertising or indecency is regulated in the case of broadcasting. Merely fighting spam or attributing the origin of fake news does not address the crude manipulation of opinion and emotions that are the source of informational, political, and civic problems. Instead, the self-regulation of AI represents a surrender to depoliticized norms in which political issues are addressed in purely legal or civic terms [32]. Yet censorship and privacy rules will never solve the bulk of integrity problems that are present in communication because they derive from processes that are not a monopoly nor a cradle of disinformation. The respect for privacy and the constitutional rights that it represents are related to the manipulation of companies, not to mass manipulation. On the other hand, the more controversial the content that AI companies regulate, the more questions about the legitimacy of regulation that do not arise for issues such as patent law or data protection will multiply. Treating politics and communications from the perspective of the liberal ideal of tech companies devoted to the production of social welfare will surely lead to ignoring the demands of other social actors who perceive negative consequences from the abandonment of a set of classifications of goods that companies do not share [33]. We must not rule out anything that can prevent the proliferation of electronic devices that identify, direct, and feed radical communities that did not exist. To continue prioritizing the construction of a highly deliberative public sphere and to implement the necessary judicial system would be more accurate. The virtues of the deliberative public sphere or the regulation of impartial institutions are values supported by the technologies developed by AI companies [34]. These values are neither eternal nor inevitable. But if what we want is a democracy based on the public control of the means of communication, it is concerning that in no government or legislative institution does the discussion arise about the ownership and direction of technologies that statically determine what human beings perceive as legitimate and acceptable consensus [35].

6.6.2 *Impact on Democracy and Society*

The search and recommendation algorithms described above are responsible for driving about 70% of the time users spend on various platforms, and for about 60% of mobile searches. They have attracted increasing evidence that they are not only challenging the traditional view of political communication, voting behavior, and public opinion, but are having real political impacts. For instance, in 2018, it was estimated that the company influenced how 1.3 billion entities across the world perceived a significant election. Other platforms manipulated a major election and many other events [36]. In a vivid example of their political influence, during a fiscal showdown in the US legislature in 2011, a single rumor tweeted by a financial blogger was picked up by the aggregators in social news sites and amplified into a rumor that the US had lost its credit rating. This led to an immediate crash in financial markets. Sources of misinformation and disinformation are indeed multiple and include social bots and cyborgs, foreign governments, technology companies, and conventional media. Ignoring internal domestic dynamics at play, social bot and cyborg technology companies have developed algorithms that maximize information diffusion, independent of the veracity of the information and the impact on democracy and users [25].

6.7 Combating Misinformation

In the long run, serious efforts to foster greater reasoning and critical thinking about the media, the incentives, and possible regulations to help platforms slow down the spread of misinformation, improved algorithms, and better-documented data could potentially hold promise. While reasoning and critical thinking are definitely crucial, education is a long and often unobservable process. For significant short-run mitigation, there needs to be more than just hashtags or news in schools. An obvious short-run solution for any audience is the more extensive training of professionals who are in constant contact with any audience and susceptible to media manipulation [37]. Educating and certifying teachers who are, in turn, able to educate their students could facilitate a significant increase in the level of awareness and expertise in the long run, while simultaneously reducing the time constant. One of the weakest links that require high resilience and that can be easily strengthened in current systems is that of the media professional. In addition, while unsupervised AI may be part of the problem, supervised AI can also be a big part of the solution. AI such as algorithms to reveal misinformation, determine which misinformation was created by AI and counter it. These machine learning algorithms can automatically detect deep fakes and source deep technologies. These algorithms can automatically detect the origin of a media file [38]. These algorithms can use metadata and source recognition to determine if data is large enough or possible to report, and in general AI trained to support journalism and help journalists and fact-checkers.

6.7.1 Fact-Checking Initiatives

The most direct way to counter misinformation is through fact-checking. The first dedicated websites appeared in the United States in 2004 with various organizations. Since 2010, and especially since 2016, their number has grown dramatically. In the European Union alone, there are more than 90 fact-checking organizations, either as individual entities or as specific verticals within larger media houses. Media and corporations are also increasingly running their own fact-checking desks. Besides the US and Western Europe, there are now many fact-checking organizations in Latin America and Asia. In 2014, five of the major fact-checking organizations united to form an autonomous network. Fact-checking organizations have guidelines for the verification of political claims and coordinate regular meetings between them in order to discuss their practices. As respected arbiters of truth or falsity in debates and electoral campaigns, they employ the attention generated by false claims to counteract their effects [39]. Research has documented the effect of politically oriented fact-checks: they are successful in correcting false information for politically uninformed people, but the evidentiary basis for their impact on those with prior political knowledge is less clear, as they are more resistant to falsifying information.

6.7.2 Media Literacy Programs

To conclude, we have to recognize that there are no silver bullets against insidious systems like deepfakes, misinformation, and online manipulation. They are the consequences of a more profound crisis of media disinformation or even "post-truth." Combating them should be, indeed, a national priority. Contrary to what one may believe, there is no single technical fix. Countering deepfake videos, images, audio, and all other forms of media manipulation ultimately requires a long-term effort by a variety of actors from diverse sectors, including, of course, the engineering and technology industries, journalism, and education at all levels [40]. Media literacy programs should aim at strengthening the public capacity to evaluate the credibility of news and information, the ability to gather, process, and use information, and the development of critical thinking and the civil discourse capacities needed to become citizens who are able to manage daily information in their personal lives and participation in society. Children, parents, and citizens, in general, should be prepared to "be behind the screen" with conscious aptitude and application. They ought to be "code aware," thus developing the media and digital competencies necessary nowadays in the always-on communities of postmodern societies [41]. The effort is immense, but the fight in the digital realm should not exempt the construction of vigorous foundations of democratic debate and its dialectic under the post-digital war, as presented in Table 6.1.

Table 6.1 AI-driven misinformation in media and politics: risks, tools, and mitigations

AI tool/ technique	Use case	Impact on media/politics	Ethical concern	Affected stakeholders	Proposed mitigation
Deepfake generation	Fabricated videos	Spread of fake political speeches	Erosion of public trust, identity misuse	Politicians, citizens	AI-based deepfake detection tools
Language models	Fake news articles	Flooding the internet with misinformation	Manipulation of public opinion	Media consumers, journalists	Fact-checking integration with content
Social media bots	Content amplification	Artificial virality of biased narratives	Lack of transparency in information spread	Voters, platforms	Bot detection and removal policies
Sentiment analysis AI	Campaign monitoring	Micro-targeting and emotional exploitation	Undue influence on voter emotions	Political parties, citizens	Ethical ad policies, transparency tools
Algorithmic curation	News feed personalization	Echo chambers, confirmation bias	Polarization and reduced critical thinking	Social media users	Algorithm transparency, diversified content
Generative text tools	Fake political commentary	Disruption of informed discourse	Misleading narratives	Public opinion, journalists	Watermarking AI-generated content
Image synthesis models	Fabricated evidence	Misleading visuals in propaganda	Visual misinformation	Law enforcement, media	Visual content verification tools
Chatbots in campaigns	Automated engagement	Illusion of public support	False representation of opinions	Political activists, voters	Disclosure of AI in political interactions
Voice cloning AI	Impersonation of leaders	Fake audio leaks, reputation damage	Consent violation, trust issues	Public figures, audience	Legal restrictions on voice cloning
Targeted disinfo ads	Political advertising	Spread of false narratives to key groups	Voter manipulation	Electoral bodies, citizens	Real-time ad screening and monitoring

6.7.3 Regulatory Approaches

In this chapter, we have highlighted the significant challenges of addressing harmful uses of artificial intelligence (AI) in media and politics. We first offered an overview of the range of problems associated with these uses, including misinformation, disinformation, data privacy concerns, and political expression harms. After that, we collated thoughtful suggestions from scholars and practitioners around the world. While there was considerable consensus regarding the gravity of the challenges and the areas in which interventions were needed, there was a variety of opinions on how to address these harms best. In this section, we consider some of the dominant themes to emerge from the literature and address the question of how various interventions should be prioritized in an imperfect policy world [42]. There is a growing number of governmental and international organizations that are attempting to address the concerns associated with AI. These have been a result of both democratic nations choosing to better regulate its use and international intergovernmental organizations coming to decisions on the common security concerns associated with AI. An analysis has identified 61 international organizations with the capability to shape rules for AI, the most capable of which, in terms of important positions or likely attention to AI, were the United Nations, the World Trade Organization, the Organization for Economic Cooperation and Development, and the International Telecommunication Union [5].

6.7.4 Current Laws and Regulations

In the European Union and its Member States, certain laws and regulations partially govern the practices of misinformation and manipulation in media and politics. In general, few rules directly address misinformation and manipulation, but animal spirits have an inherent ability to cause changes in this domain. Data protection, privacy, consumer, competition, advertising, and political or electoral laws and regulations are involved, albeit from different angles and with different scopes and degrees of stringency and binding force, on both online and traditional analog media [2]. As we discuss here, the more private-oriented legal regimes are much stricter and more demanding than the politically prejudiced and much laxer public law framework. Within the disjointed legalistic ecosystem, it is essential to map, demystify, and disclose the existing laws and regulations related to misinformation and manipulation in the context of AI in media and politics. In a sequential way through varied legal systems, we will start by critically addressing those whose first recipients of the regulation are the big commercial stakeholders in the ecosystem, namely data-driven digital platforms as follows: data protection and online services in general; privacy and individual personality rights; consumer law and the digital single market; competition, advertising, and strategic commercial communications; security, electoral matters, and political and public interests [3].

6.7.5 Proposed Policy Changes

Identifying safe and fair ways to alter AI systems is crucial to addressing its role in misinformation. Since AIs are costly to produce, regulate, and often control data that is proprietary, policy is likely to be the only practical way to induce such changes at scale. We propose the following changes that rest on our understanding of current and near-term AI challenges:

(i) Narrow the objectives AIs pursue. Current systems have errors that change when politicians or journalists act differently. Examine these through counterfactuals: what are the NLP system's decisions (or errors) if something else occurred? Encourage corporations to research methods for limiting this dependence. Another promising approach is to train the AIs to plan at a higher level of abstraction, where research has seen some success.

(ii) Publish and use assessments. To better understand why current systems produce their outputs in any specific context, require tech corporations to publish such assessments. And perhaps more importantly, make such assessments prerequisites for using AI when politically relevant.

(iii) Continue to support independent audit efforts. Corporations and academics are starting to align their interests and to execute such investigations. Govern these efforts so that companies remain liable. Liability is an equalizer: it aligns incentives for all parties without making special treatment for any.

(iv) Encourage political experiments. Political actors that collect and analyze personal information about citizens are often very powerful, and under the status quo, firms have a big advantage in using AI. However, advertising is only one reason for these experiments. Encourage other, less unregulated uses, especially when it comes to the fabrication of fake images or audio. Since doing so collects data that is often more protected, the natural ambition would be for political actors to abandon their experiments [41]. Excusing a lot of behavior would solve this problem, along with some other privacy issues too.

(v) Discourage and regulate media oligarchy. While accountability could break some aspects of these oligarchies, doing so is not easy. Concentrated media interests and policy decisions that affect these interests have always been a feature of democratic societies. But existing large media businesses greatly amplify the dangers that AI poses, especially those influencing content popularity.

(vi) Facilitate the emergence of new voices and promote public interest journalism, possibly through external media endowment, though of course politicians will often use that to their advantage [38]. Monitor that use and continue to support independent depolarization media efforts.

6.8 Future Trends in AI and Misinformation

Machines are becoming more than just machines and increasingly occupy the space between humans and computers, affecting culture and human expression in multiple ways. AI in various forms will continue this trajectory while interacting with misinformation at every turn. The described impact of certain algorithms could have been made about many others, and in less than a year, the same would be said for newer models [43]. If demonstrated abuses played a meaningful role in slowing down the dissemination of certain technologies, it is an open question what society would need to do to slow the genie this round. However, there are relatively easy projection techniques to forecast where AI capabilities will be in a few years [42]. Trying to restrict legitimate AI for misuse of misinformation-related tasks and keep a thin layer of 'armor' to avoid misinformers might not have tops. However, going beyond these near-term forecasts to envision how future advances in AI and related technology could travel and interact with what the misinformers have been better requires deeper thought. A set of conversations is necessary—not only among professionals in the field but also between the AI community and the broader public it serves. In this survey, given the critical nature of impending threats of misinformation, we have discussed a large range of tasks in media and politics, offering a detailed exploration of the dark edges of AI. However, misinformation is not the only such threat to democracy and society, as the apparent use of AI in targeted harassment or deep fakes shows [3]. As a first step towards these conversations, in the following pages, we set the basis for a distinct effort in grading not only what AI tools are capable of (when hard enough) but also how these advancements may be considered according to their long-term impact on society.

6.8.1 Advancements in AI Technology

Developments in AI technology have enabled remarkable progress in the exciting applications described above. Fueling tech hype, the acceleration of machine learning subfields termed deep learning and reinforcement learning is prominent. Both rely on large neural network architectures, feature highly flexible design options on a strong math-based footing, and excel in areas such as pattern recognition and statistical processing. The key innovation in deep learning that dramatically increased its popularity is the empirical success of so-called convolutional neural networks. Convolutional neural networks are deep learning architectures engineered to handle 2D or 3D input data that exhibit a locally translated 2D-to-1D feature mapping in an organized way [42]. Shallow feature extractors are replicated and then sparsely integrated in a deeper layer to produce invariant and order-preserving activation patterns.

Unlike traditional feature engineering employed in most machine learning applications, the learning algorithms in deep learning extract useful feature representations from pixel raw data. Convolutional neural network-based feature hierarchies have been demonstrated to be equivalent to carefully handcrafted state-of-the-art image feature designs in benchmarking datasets and to the design of neurons in the early stages of the mammalian visual cortex. While the progress in supervised learning based on convolutional neural networks has outperformed designs based purely on feature engineering, it is the dominance of big data and deep learning over the traditional approach that draws attention [38]. Small-sized single-neuron integrated CPU chips that are now economically available use the architectures behind large-scale cloud processing services, together with immense raw processing power in the form of programmable graphical processing units, have been essential to securing these successes.

6.8.2 Predictions for Media and Politics

AI and deep learning, in particular, are going to continue to shape the way we consume media and interface with one another and with society in the decades to come. It might be troubling, then, to consider the dark potential for the technology in the context of media and politics. The term "personalized propaganda" might become more potent in that context. Mercifully, such nefarious applications are limited by the nature of consumer demand: ratings are still overwhelmingly determined by consumer interest in authentic, high-quality, ad-free reporting and programming that protects the public interest. But, as various scandals and controversies have illustrated, commercial and political interests can bankroll disinformation [25]. AI news has no fury like natural language processing news meets big data news. As we learned from the flood, the high-dimensional complexities of real news are often poorly approximated by flat data. Creativity and conscious modeling are frequently required. That being said, the forecast calls for a never-ending bulk increase of AI applications to both improve reporting accuracy, breadth, depth, and scale and tailor content; likewise, we anticipate a steady increase in the development and refinement of detection software on both the news production and news consumption ends of its applications [16, 29]. AI could also treat humanity with beneficence via curatorial and personal shopper capacities for opinion development, helping to reduce the echo chamber of affirmation bias.

6.9 Conclusion

In the introduction, we discussed several reasons why the current worrying situation does not yet amount to a full-blown crisis, and we revisited a number of these points once more in the conclusion. The boundary between persistent difficulties and outright crisis remains difficult to draw, especially because much of the measure of the impact of misinformation on society depends on the capacity of society to generate mechanisms that counteract misinformation's worst effects. Several reasons to be optimistic remain valid, and additional ones have probably become salient throughout the crisis. As a result, we may be somewhat less worried about the current situation than when initiating this inquiry, including at the point when misinformation in connection with the pandemic reached what might be termed epic proportions. But we also encounter reasons to be more worried than when writing the introduction. One intriguing feature of the current relationship between society and misinformation is that the crisis feeds back on the structural challenges society has to tackle. Social platforms are incentivized to broaden the reach of trustworthy sources while also making their services a less fertile ground for the distribution of misinformation. A number of recent fact-checking initiatives and the behavior of traditional news media provide hope. Most importantly, resources to invest in coping with this situation are growing at a time when information overload is increasing. The worrisome potential for disenchantment with science, expertise, and the role of traditional media may, in the end, initiate a rather forceful move in the right direction—provided we realize that we must not go too far before overcooking it all. In that sense, a momentary convergence can be observed between the status quo misbalances and the more bound-to-stay concerns we identified at the beginning of the earlier section.

References

1. Yildirim A, Yolcu E. Sahte Ne Kadar Derin? Derin Sahte (Deepfake) Kavramının İzini Youtube Üzerinden Sürmek. Elektronik Cumhuriyet İletişim Dergisi. 2022;4(1).
2. Callahan WA. Citizen AI: Warrior, jester, and middleman. J. Asian Stud. 2014;73(4).
3. Shim JS, Lee Y, Ahn H. A link2vec-based fake news detection model using web search results. Expert. Syst. Appl. 2021;184.
4. Diez-Gracia A, Sánchez-García P, Martín-Román J. Disintermediation and disinformation as a political strategy: use of AI to analyze fake news as Trump's rhetorical resource on Twitter. Prof. Inform. 2023;32(5).
5. Zekos GI. AI and politics. In: Contributions to political science. 2022.
6. Shafik W. Human-computer interaction (HCI) technologies in socially-enabled artificial intelligence. In: Future of digital technology and AI in social sectors. IGI Global;2025. p. 121–50.
7. Elliott A. The routledge social science handbook of AI. 2021.
8. Shafik W. Generative AI for social good and sustainable development. In: Generative AI: current trends and applications. Springer;2024. p. 185–217.
9. Shafik W. Deep learning impacts in the field of artificial intelligence. In: Deep learning concepts in operations research [Internet]. New York: Auerbach Publications;2024. p. 9–26. https://doi.org/10.1201/9781003433309-2
10. Shafik W, Kalinaki K, Fahim KE, Adam M. Safeguarding data privacy and security in federated learning systems. In: Federated deep learning for healthcare [Internet]. Boca Raton: CRC Press;2024. p. 170–90. https://doi.org/10.1201/9781032694870-13
11. Strafella G, Berg D. "Twitter Bodhisattva": Ai Weiwei's media politics. Asian Stud. Rev. 2015;39(1).
12. Shafik W. Generative artificial intelligence for social good and sustainable development. In: Generative AI: disruptive technologies for innovative applications. 2025. p. 153–85.
13. Shikhar, Teckchandani J. AI in international politics. Int. J. Res. Appl. Sci. Eng. Technol. 2024;12(3).

14. Köstler L, Ossewaarde R. The making of AI society: AI futures frames in German political and media discourses. AI Soc. 2022;37(1).
15. Stahl BC, Schroeder D, Rodrigues R. Ethics of Artificial Intelligence: case studies and options for addressing ethical challenges (excerpt). Ekonomicheskaya Sotsiol. 2024;25(1).
16. Parviainen J, Coeckelbergh M. The political choreography of the Sophia robot: beyond robot rights and citizenship to political performances for the social robotics market. AI Soc. 2021;36(3).
17. Carral U, Elías C. Application of AI tools as methodology for the analysis of toxicity in social media: a case study of Spanish politics on Twitter. Rev. Latina de Commun. Soc. 2024;2024(82).
18. Dutton T. Medium. 2018. An overview of national AI strategies—politics + AI—medium
19. Mega RAYS. Countering democratic disruption amid the disinformation phenomenon through Artificial Intelligence (AI) in public sector. J. Manajemen Pelayanan Publ. 2023;7(1).
20. Westermann C, Gupta T. Turning queries into questions: For a plurality of perspectives in the age of AI and other frameworks with limited (mind)sets, vol. 21. Technoetic Arts; 2003.
21. Shafik W. Generative adversarial networks: security, privacy, and ethical considerations. In: Generative Artificial Intelligence (AI) approaches for industrial applications. Springer;2025. p. 93–117.
22. Jun Y, Craig A, Shafik W, Sharif L. Artificial Intelligence application in cybersecurity and cyberdefense, vol. 2021. Wireless Communications and Mobile Computing; 2021.
23. Shafik W. Cyber attacker profiling and cyberbullying overview. In: Cyber space and outer space security [Internet]. New York: River Publishers; 2024. p. 125–49. https://doi.org/10.1201/978 1003558118-5
24. Shafik W. Toward a more ethical future of artificial intelligence and data science. In: The ethical frontier of AI and data analysis [Internet]. IGI Global; 2024. p. 362–88. https://doi.org/ 10.4018/979-8-3693-2964-1.ch022
25. Bakir V, McStay A. Optimising emotions, incubating falsehoods: How to protect the global civic body from disinformation and misinformation. 2023.
26. Alaziz SN, Alshowiman AA, Albayati B, El-Bagoury A al AH, Shafik W. Clustering of COVID-19 multi-time series-based K-means and PCA with forecasting. Int. J. Data Warehousing Min. 2023;19(3).
27. Mantello P, Ho MT, Nguyen MH, Vuong QH. Machines that feel: behavioral determinants of attitude towards affect recognition technology—upgrading technology acceptance theory with the mindsponge model. Human. Soc. Sci. Commun. 2023;10(1).
28. Chen Q, Srivastava G, Parizi RM, Aloqaily M, Ridhawi I Al. An incentive-aware blockchain-based solution for internet of fake media things. Inf. Process Manag. 2020;57(6).
29. Ascott T. Microfake: How small-scale deepfakes can undermine society. J. Dig. Media Policy. 2020;11(2).
30. Grondin D, Hogue S. Person of interest as media technology of surveillance: a cautionary tale for the future of the national security state with diegetic big data surveillance, algorithmic security, and artificial intelligence. Telev. New Media. 2024;25(4).
31. Burema D, Debowski-Weimann N, Von Janowski A, Grabowski J, Maftei M, Jacobs M, et al. A sector-based approach to AI ethics: understanding ethical issues of AI-related incidents within their sectoral context. In: AIES 2023–Proceedings of the 2023 AAAI/ACM conference on AI, ethics, and society. 2023.
32. Ghosh S, Ekbal A, Bhattacharyya P, Saha T, Kumar A, Srivastava S. SEHC: a benchmark setup to identify online hate speech in english. IEEE Trans. Comput. Soc. Syst. 2023;10(2).
33. Lyons M. Excavating "Excavating AI": the elephant in the gallery. SSRN Electron. J. 2021.
34. Shafik W. Artificial Intelligence and machine learning with cyber ethics for the future world. In: Future communication systems using artificial intelligence, internet of things and data science [Internet]. Boca Raton: CRC Press;2024. p. 110–30. https://doi.org/10.1201/9781032648309-9
35. Klipphahn-Karge M, Koster AK, dos Santos Bruss SM. Queer reflections on AI: uncertain intelligences. 2023.
36. Astobiza AM. Do people believe that machines have minds and free will? Empirical evidence on mind perception and autonomy in machines. AI and Ethics. 2023;

37. Shafik W. Ethical and legal considerations in artificial intelligence. In: AI-enabled threat intelligence and cyber risk assessment. 2025. p. 90.
38. De Togni G. Staging the Robot: performing techno-politics of innovation for care robotics in Japan. East Asian Sci. Technol. Soc. 2024;18(2).
39. Pattison A, Cipolli W, Marichal J, Cherniakov C. Fracking Twitter: utilizing machine learning and natural language processing tools for identifying coalition and causal narratives. Polit. Policy. 2023;51(5).
40. Eacho D. Performativity without theatricality: experiments at the limit of staging AI. Theatre Perform. Des. 2023;9(1–2).
41. Ji J, Hu T, Chen Z, Zhu M. Exploring the climate change discourse on Chinese social media and the role of social bots. Asian J. Commun. 2024;34(1).
42. Trandabat D, Gifu D. Discriminating AI-generated fake news. In: Procedia computer science. 2023.
43. Shafik W, Zakari RY, Kalinaki K. Ethical and privacy concerns in bioinformatics and cyber-physical systems integration in healthcare. In: AI-driven personalized healthcare solutions. IGI Global Scientific Publishing; 2025. p. 333–64.

Chapter 7
The "Black Box" Problem: Lack of Transparency in AI Decision-Making

7.1 Introduction

The term "AI decision-making" makes it sound like AI is creating models as motivated agents sitting at the control panel. However, as we know, machines do not have minds. Latent decisions, or "decisions to act," are hardcoded into their design. That is, humans pre-make many decisions about design or action to reduce the capability of the machines' latent, mechanical decisions. Understanding AI decisions is ultimately about understanding humans and the complexity of the decisions we design or cause a model to make. Pre-permutation feature importance metrics typically measure the inter-model variability of a prediction for a single instance, with the feature value being masked [1]. The average variability across other instances produces its measurement. Because of the masking process, influenced features have zero or marginal influence depending on whether that instance has that feature value. This method is at odds with the common sense that a feature's importance is measured by how much of a difference it is attributed to decisions across all instances. Despite the method being misaligned with its goal, influential features are often the main suspects in being of higher importance than they actually are [2].

7.1.1 *What is AI in Decision Perspective?*

AI technologies have been around since the 1960s and have allowed us to do certain things that mimic human intelligence, such as discovering proofs of mathematics or playing chess. There are many definitions of AI; depending on the particular application, one might favor a narrow definition for regulatory reasons that distinguishes between truly autonomous systems and AI supporting or augmenting human decision-making capabilities. We describe "autonomous AI" as the system "where

W. Shafik, *The Dark Side of AI*, https://doi.org/10.1007/978-3-032-09130-7_7

such a system can make independent decisions or facilitate a process to reach conclusions, including, but not limited to, any vision systems, robotics, integrative systems, predictive systems, or cognitive systems that use or enable automated processes, computational processing, or enact actions that are driven by data, to sense, evaluate, reason, learn, or facilitate decision-making in any domain of activity, in real-time or later" [3]. The distinctions between narrow and general definitions of AI are important, as they allow us to differentiate more transparent, cooperative AI from more opaque AI that sets its own goals and may refuse requests to focus its faculties on higher-level goals. Whether the specific machine that enacts AI is regular or autonomous is less interesting for our immediate purposes than the distinction between narrowly intelligent machines and general machine intelligence. The term "black-box" is often used in financial analysis and cyber-physical security to mean that the operational status of a system is opaque to non-observers and is therefore difficult to analyze, even if the statistical dynamics of individual aspects of the system are well characterized and relatively transparent [4]. Frameworks to transparently analyze complex systems are well-developed, as are frameworks to transparently analyze the mode of operation and the potential static and dynamic security vulnerabilities of systems built with software.

7.1.2 How AI Makes Decisions

A less obvious question is how we judge AI's decisions. Although AI often works in unpredictable ways, over time, its growing ramifications require a measure of predictability. A key aspect of human decision-making is that, often, we are able to anticipate with some confidence the outcomes of our actions. By contrast, once certain types of AI have been trained, it is often difficult for us to reverse-engineer the processes by which they make decisions. Similarly, after training, AI models are rarely able to revise their decision processes. Even holding constant the complexity of AI, its low decision-making transparency generates irreducible problems [3]. First, without transparency, it is difficult to predict or understand the outcomes that AI is liable to generate. Second, people usually do not accept decisions that they are asked to accept merely because a 'black box' tells them to through different processes of decision-making as presented in Fig. 7.1.

The failure to interrogate AI with respect to how decisions are made is independent from any lack of AI formalism. The nature of many AI analyses can be quite transparent: what AI 'sees' is often much clearer than what a human observer is able to see. Indeed, much of the success of AI generally, and deep learning in particular, rests with this insight, achieved through the application of relatively simple statistical methods often applied against massive datasets and vast computational resources and then tested against various standards of performance. Nonetheless, understanding how AI 'sees' something is not nearly the same as understanding how it acts based on that perception [5]. The apparent separateness of the AI act may be one cause of the strong incentive AI developers have to employ poorly interpretable,

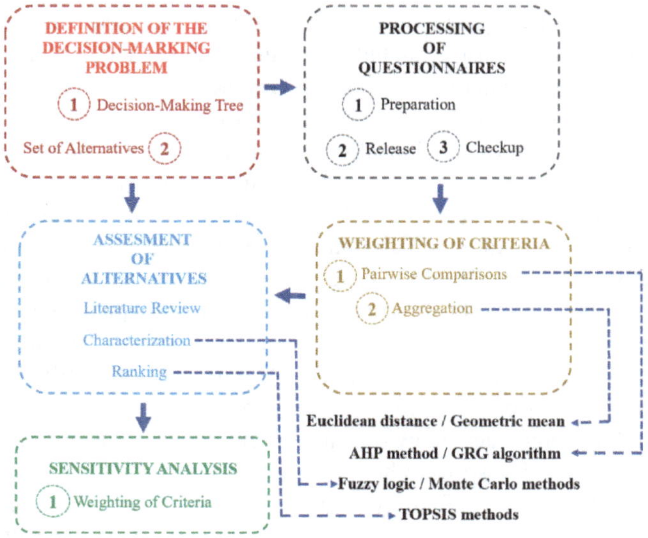

Fig. 7.1 Decision-making processes

complex machineries for decision-making: the perception of decision-making separateness might lead to a less demanding set of interpretative standards and differing normative assumptions. Establishing relevant accountability, however, requires that researchers, developers, and users consider both performance and the underlying AI structures that give rise to particular decisions [6].

7.2 The Black Box Phenomenon

As shown in the previous discussion, the operational logics of AI, including input coding, data collation, and result analysis, cannot be completely understood or verified, and AI working mechanisms act like "black boxes." As AI works and accomplishes activities with these mysterious "internal" logics, people gradually believe in and rely on the authority and fairness of AI opinions—AI also obtains power and control invisibly. However, the black box features could hide the potential for irrational operation, incorrect influencing, and unfairness caused by incorrect coding, incomplete data sources, or biased data. Notably, some existing AI tools' behaviors in equality, transparency, and objectivity have been challenged. At the same time, compared with the precise and objective decisions reached by AI, human decisions concurrently produced by AI tend to be accepted [7]. Although AI reduces human bias and prejudice, its legendary rational, objective opinions can become human biases and prejudices, influencing decision-making, justice, and fairness in human society.

The occurrence of the black box might have both technical and non-technical reasons. First, technical reasons include computational capacity and data. With the development of cloud computing and big data techniques, we can calculate and learn from much data and deal with complex structures of data precisely and quickly at lower costs [8]. Furthermore, the development of AI techniques addresses the black box problem related to the uncertainty of data structures and the collation of data by learning from the positive and negative samples of massive datasets. However, the result of those techniques is a combination of parameter information that is so complex that we cannot understand it. Second, non-technical "mysterious" reasons might describe the possible underlying social, ethical, and epistemological reasons originating from "moral value" and dependent on "moral value." The ethical, cultural, and nurturing value effects influence the design principles of AI [9]. The technical marker of AI derives from and represents the overlapping values of contemporary society, including democracy, individual autonomy, equality, justice, social welfare, and public safety. Unlike specific human uniform standards, those factors exhibit characteristics of overlap, conflict, dynamics, and intersection—designed AI cannot balance and satisfy every value simultaneously. Despite the extensive discussion of concepts such as "safety, security, accountability, transparency, traceability, corruptibility, fairness, vulnerability, quality, and reliability of the created technology," AI tools do not possess external indicators to ensure their behavior is trustworthy [10]. To answer the question "Is this trust proper?" is simultaneously neither technical and not neutral.

7.2.1 Definition of the Black Box

Until the mid-1990s, AI was characterized as knowledge processing. It processed what the expert knew. Often, it involved a specialist with a specific task, not just the average person. When questions were asked about why the expert thought a certain way or what rules were applied, the answers were given by the experts who knew how the knowledge was encoded into the system. This environment is most likely known as the "white box," where technology is fully transparent and understandable, in contrast to the technology structure, which is considered a black box or transparent block that gives no insight into how it works. However, the perception of AI started to change after the creation of "algorithmic" deduction and inductive machine learning, such as statistical class sensing and neural networks using large collections of cases in the 1980s and 1990s [11]. The original expert knowledge, domain constriction, or administration educated and characterized the "specialist" for AI in a moderately imperfect human task. When questions were asked about why the expert thought in a certain manner or what policies were applied, the answer was given by the "expert," who knew how the knowledge was encoded into a computer system. This kind of technology is known as the "white box." It is the event that the system operates entirely with expert knowledge. Software transparency is not an assumed characteristic of any program. The clarity of the software is determined by the expert's capacity and willingness to clarify how to determine the solution and the applications that offer access to the program's internal operation [12]. Various black box systems are used by people, but are used to accept actions that are beyond their control in several instances.

7.2.2 Historical Context

In this section we review the history of black-box problems, mainly those related to non-AI-based safety critical systems, in particular from the perspective of opacity. The black-box problem is not, of course, a monolithic entity. Safety-critical systems have long been a component of, for example, health monitoring systems for humans. Many of these older designs do exhibit the type of system-level transparency that AI systems do not [13]. For example, in older safety-critical designs, the watchkeeping health monitoring system of a warship engined room generally had any other systems that monitor the propulsion control system. For example, the design of a third turret was a premature blast of one of these outdated health monitoring systems. In 1943, one of the most powerful battleships ever built had problems with the three large guns in each of its three turrets, which had been found and eliminated [9]. Then, while the watch was underway, a gun fired, and the resulting explosion killed 47 crewmembers. In the investigation that followed, it was discovered that in the watch's standard operating procedure, there was an unwritten, but well-known modification that had gradually come into use. Since kitchen personnel needed to steam large amounts of food for the watch's personnel, the battleship allowed them to order their food in order to do so. On occasions such as this, the bridging station would be in the standard bridging light system, used to alert the operators if there was going to be a firing. On the specific day of the incident, the light began to be realized too late for it to be able to evacuate the firearm system of the turrets and provide the turret with a good vantage point [6]. When the light finally appeared, the light was disregarded, as the gun was still authorized to fire, and the explosion occurred almost immediately.

7.3 Implications of Lack of Transparency

The lack of transparency in AI is not merely a practical challenge but can create significant legal and ethical obstacles. Transparency is crucial in many legal and ethical arenas, such as business regulations, tort liability, equal protection rights, administrative law requirements, and doctrinal constraints on criminal law. These constraints are particularly significant in situations where these decision-making systems are incorporated into processes that have significant impacts, including criminal justice processes and systems subject to constraints. It also raises concerns about potential issues, as cases interpreting official and unofficial promotions of racially biased traffic enforcement have been thrown out because of these issues [5]. With or without a lack of transparency, AI decision-making tools may add significant and unpredictable issues to many legal and ethical areas. The lack of transparency, which all decision-making tools in statistical analysis face, is particularly concerning in AI decision-making tools given their complexity and changing dynamics. This difficulty in emphasizing the importance of transparency demonstrates not only the activities but also suggests the need for additional substantive regulations to provide insider and outsider information. While regulators often pass regulations to ensure substantive evaluation, regulations are largely taken in these results as safe behavior from autonomous vehicle research by companies developing fair AI decision-making tools [4].

7.3.1 Ethical Concerns

Now that we have different AI modalities in our lives and are using online tools for our studies and different technologies for our entertainment, it is high time for us to discuss the values we must adhere to steer a clear course. Notably, AI's ability to mimic human intelligence, assess patterns, forecast long-term trends, and handle substantial data raises different ethical issues, as there is no parallel method for achieving information and decision support. Contrast it with the ethical requirements for large datasets, predictable and stable decision-making processes, the identification of precise values to incorporate into decision-making, and a clear understanding of who is responsible [14]. As such, no single answer suffices, and lawyers, computer scientists, data scientists, ethicists, and social scientists should collaborate to answer ethical questions. Upon closer examination, it becomes apparent that a lack of transparency in AI algorithms facilitates the development of unaccountable automated decision-making systems. AI has also made rapid progress in such complex tasks as face recognition and language understanding in the past few years. Indeed, these rapidly expanding systems have raised new concerns. AI can often achieve higher performance than humans in some tasks, but its decision-making process remains a black box [15, 16].

How can we trust the conclusion if we don't understand the method by which it was reached? Furthermore, how can we be assured that the AI is deciding by complying with ethical constraints in the absence of any transparency? The AI profession echoes these concerns and has formulated ethical standards to recommend autonomous decision-making to be suitable for public use [17]. However, while realizing the potential benefits of knowingly incorporating ethical requirements, we must keep in mind that the necessity to adhere to those guidelines of autonomy is an explicit recognition that truly autonomous decision-making is a difficult challenge [18]. Put, incorporating the existing ethical principles into practices via software is not a sufficient or warranted trade-off between transparency and other variables such as efficiency and privacy. Then how can any black-box decision-making software be entrusted? Even if an organization decides in good confidence and has no vested interests, can we definitively affirm that the organizational values are actually reflected in the decision-making process? In the interests of unraveling these issues and exploring answers to these difficult questions, i.e., those resolution routes that encourage transparent software, this text is designed to provide an underlying rationale for the understanding of 'ethical compliance software' [19].

7.3.2 Impact on Trust

Rachel Cummings, who researches non-traditional questions in data management, including the interpretability of 'black box' models and testing of learned models to ensure levels of fairness and accountability, is already seeing companies asking about how interpretable a model is. She says, "There's increased awareness of this and decreasing trust in models that are not interpretable. And if that's what's happening for people who are working in the industry, then how is that going to be reflected among consumers? Just as the public is seeing how companies are using their data, they are becoming more cognizant of when an AI model is used on them. This impacts not only a company's bottom line but also trust. If a company says, 'Your loan was denied based on a machine learning model,' you're left wondering [20], 'Why?' and as a consumer, how would the bank prove to you that you were denied for acceptable, non-discriminatory reasons? This is a big question." Even if no law mandates an explanation, a company might still want to consider the ramifications of hiding such decisions and eroding trust. Especially since trust in, and public support for, AI technologies will be essential if their full benefits are to be enjoyed. With trust in both companies and government remaining low, AI will face challenges with public perception of these systems, trust in said systems, and issues of accountability in decision-making. With trust in corporate America at an all-time low, firms can help address this situation by promising that their AI judgments are interpretable [21]. The burden of proof is on those who deploy machine learning; show us that these decisions are safe; in other words, install a window into this black box.

7.3.3 Legal Ramifications

Companies selling products using AI systems need to worry about liability when the AI makes decisions that result in consumer harm. With a lack of transparency, company leadership might not even know how or why it might have happened, let alone how to fix it. Product liability statutes may hold vendors liable when a substandard product design harms a user. However, applying product liability laws to AI systems is complicated, and their application in SDGs [22]. Due to their basic capabilities, employees, customers, and vendors of an AI firm do not view the AI system as a typical product design held to liability laws. In traditional product liability cases, whether the product possessed safety defects or was in a defective condition when it left the seller's control was the typical inquiry. But AI systems continuously iterate and learn from observing new data over time [23]. A company may improve its system over time, so delays in recognizing and correcting potential safety or quality concerns may not only result in harm but could also subject the vendor to liability if a court does not understand the technical factors that led to a decision. AI decision-making, often mainly reliant on large databases and complex calculations, can impact the meaning of these requirements and the application of reliability principles in product liability law. To apply traditional product liability models to AI systems, the system must meaningfully render an AI system as a "product" that "defects" within the realm of established jurisprudence, even while AI systems differ substantially from traditional products, and their risks might arise from reasons categorically distinct from traditional product failures [24].

7.4 Case Studies

The lack of transparency in AI decision-making can often lead to undesirable or even harmful outcomes. We now cover some notable instances. It is important to note that the problem is not a new one. Ever since filtering engines were first deployed, it was found that inappropriate material generated by one user could, through the internal mechanics of the recommendation and learning engine, influence and promote equally inappropriate material elsewhere, despite the domain setting that indicated interest based on likelihood of requests and resulting content. Significant efforts were made to mask these harmful effects, with laws formally laying the foundation for filtering engines being extricated from conventional notions of one's responsibility for third-party content and repurposing consumer expectations when construed in any other form [25].

The general problem has persisted throughout—the lack of transparency coupled with the ability for AI-driven applications, through their connections to their user population, to inadvertently amplify the inputs being provided has spawned notable concerns about copyright violations and counterfeit items in search engine returns, therapeutic escalation in high volume hiring systems, health misinformation in recommendation and learning engines, and the amplification of age, race, and disability discrimination in the distribution of medical resources via commercial healthcare algorithms. Not surprisingly, there have been numerous public sentiment-based movements that sought to limit or constrain the opacity of these services, such as those directed towards the transparency of legislative outputs, transparency in the operation of recommendation and learning engines, and broad AI algorithmic transparency [26]. In each of these instances, the reaction set forth has not been so much about addressing the black box problem, but about addressing the adverse implications of the black box problem. Unfortunately, this reaction is not always proper. With the possible exception of the CRM-specific instances, many of the proposed reforms are not likely to significantly increase the safety of practical systems or respect the privacy of users; the manner in which tools have been used or are planned to be deployed has more to do with the size of the losses at stake and the type of actor involved [27].

7.4.1 Healthcare AI Systems

Healthcare AI currently encompasses several major applications, including disease detection and classification, image and dataset labeling, dental applications, hospital management, mental health care, and drug discovery. Among the major methods in the domain are convolutional neural networks, recurrent neural networks, capsule networks, generative adversarial networks, and autoencoders. Convolutional neural networks predominate and can be divided in turn into models optimized to perform classification tasks and others optimized for segmentation tasks [28]. However, in terms of application, convolutional neural networks are increasingly used for solving a number of issues in the medical field, including early cancer detection, diagnosis, prediction of heart failure and coronary artery disease, and diabetic foot screening. They can analyze digital X-rays to diagnose tuberculosis at an early stage and categorize histology images. Several recent studies also propose XAI solutions. An XAI approach allows AI to analyze patients' medical history and identify suitable treatments. An approach enables AI to extract actionable clinical information from blood sample images to enable leukemia detection. A multimodal AI platform supports tinnitus understanding by extracting various attributing factors in a given tinnitus condition [29]. Medical imaging XAI techniques explain not only the classification decision but also highlight visual features related to the learned difference between positive and negative classes, using pre-trained class-discriminative features.

7.4.2 Financial Algorithms

Recent events have brought public attention to the degree to which our economy and the functioning of financial markets depend on algorithms, as well as the degree to which these algorithms are legally and commercially protected as trade secret black boxes exempted from liability for their outcomes. Algorithms are currently used across most industries to manage decision-making processes. The financial sector is no different, as they are often employed to make trading decisions or design new financial products. The ability to leverage big data sources along with larger computational capacities has given new prominence to some machine learning algorithms employed in the financial realm [30]. Algorithmic trading makes use of automated pre-programmed trading instructions to carry out trading in several different instruments in real time. Machine learning has been proposed for detecting and interpreting trading signals. Since financial data is heavily relied upon and is often noisy, this is expected to change, making such approaches feasible for simple non-learning models. The efficacy of machine learning in these applications in the financial market requires that two conditions be met: the first is that the chosen model is accurate enough so that the signals obtained from it will be reliable [18].

7.4.3 Autonomous Vehicles

Many autonomous cars currently rely on algorithms or approaches that fail to provide sufficient transparency without the utilization of additional techniques. Autonomous vehicles, or self-driving cars, live in the physical world and, as such, situational awareness is very much based on what can be perceived through various sensors, such as cameras, lidars, radars, etc. These sensors generate high-dimensional data that can be quite difficult to interpret even by humans. Depending on the driving mode, the internal perception of the vehicle might involve object classification, object detection, semantics and instance segmentation, tracking, motion estimation, and a number of other tasks. From external points of view, this means that the potential risk associated with being on the same road with these types of agents may or may not be clear [19]. Besides perception, the vehicle must be able to make tactical decisions that result in safe and efficient behavior. Although many designs involve simplifying the operational complexity of the system by specifying operational bounds, freedom may arise from the perception module. Therein lies the complexity of current models of interpretability, transparency, and explainability. In particular, an interesting aspect of driverless cars is mischief; in ideal conditions, models will train on benign states, but may be faced with adversarial examples in the real world. The unbounded nature of human and natural environments means an unbounded number of system input scenarios, which in pessimistic cases could include adversarial driving conditions [3]. To the driver or passenger of either the vehicle or adjacent cars, understanding the model's competency is crucial for its acceptance. Given the potential of AI to significantly improve road safety, it is important to guarantee that autonomous vehicle performance is contemporary with state-of-the-art.

7.5 Current Approaches to Transparency

Transparency has been a priority research area in many other machine learning subdisciplines, which have needed to ensure fairness, incentives, and compliance. Many leading voices in robotics and AI development were exposed to significant concerns about a lack of transparency. The need for increased transparency and the extent to which decisions of AI may be explainable under tests is being recognized. The area's economic significance is raising these issues in terms of practical applications. Some scholars argue that unexposed information deviates from privacy and ethics guidelines for responsible AI and does not meet the trustworthiness, accountability, or legitimate needs and rights of the holders of AI artifacts [4]. If it cannot account for their decisions, a person can attribute unfairness to an AI system and may not modify it. These legal guidelines go beyond merely requiring the transparency of AI processes; they point towards something that individuals can assure, at least in an adverse environment, that an AI agent's decisions and considerations are transparent. Legal guidelines emphasize the need to clarify AI's decision-making processes and the level of human contribution they entail. This requires the accessibility and interpretability of the algorithm to a degree necessary to ensure the appropriate degree of influence imposed on these decisions and assessments [2].

7.5.1 *Explainable AI (XAI)*

Explainable AI (XAI) refers to methods that make AI model decision-making transparent to humans. While there is no consensus on which XAI visualization technique is the best, a copious amount of research has been and is continuing to be conducted to solve problems associated with the Black Box problem. In its most general form, transparency can hinge on a model being interpretable, explainable, or providing something interpretable or explainable; several interpretations and definitions of interpretability and explainability in AI models exist, and it can depend on ontology, epistemology, the correct theory of reference, or model-bound properties [6]. Black Box models may have varied degrees of transparency and trust, and may or may not require different levels of Explainable AI. Although the full range of XAI techniques can fill gaps for degrees of explainability and interpretability required for models, many client organizations and people with various stakeholder groups cite reasons not to require explanation or interpretability as part of the Black Box model development process.

The reasons are that, in addition to explainability, many data-holding enterprises, data users, and the AI science and engineering vendor industries either lack the skills or expertise to construct proper requirements for interpretability and explanation design, lack examples of effective development processes that work from requirements for interpretability and explanation design, or lack the training data to generate effective explanations or interpretations. The goal of XAI techniques is to enhance human cognitive capabilities, without bias, to provide AI transparency, to enable human credibility through understanding, to enable ethical decision-making, to show AI's assessment and rendering of data, information, and knowledge, to provide reproducible results, and to correct cognitive lapses [7]. Cognitive lack of understanding can come from sources such as knowledge gaps, underconstrained knowledge, accuracy or precision flaws, misdirection, or attentional defects.

7.5.2 *Model Interpretability Techniques*

Model interpretation techniques seek to understand what a deep learning model has learned, thereby understanding which features influence the outcome of the decision

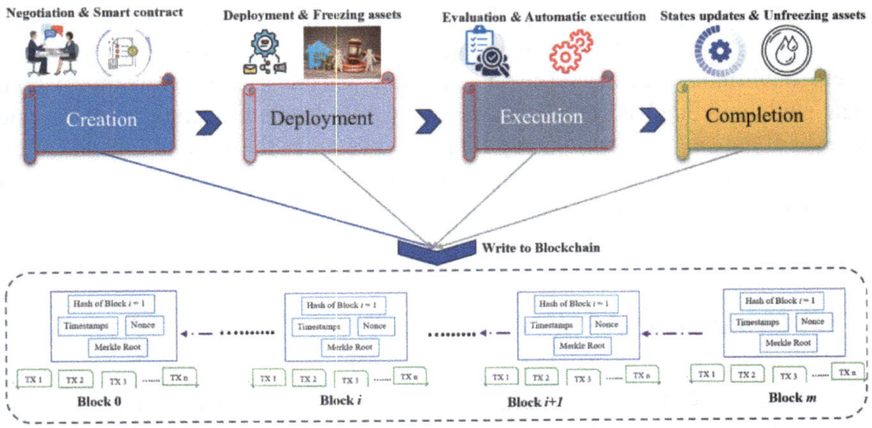

Fig. 7.2 Blockchain application for social good

process. As the model parameters can no longer be intuitively understood as is the case with shallow learning models, model interpretation techniques aim to provide insight into the decision-making models. From a business process perspective, understanding model decisions is important for evaluating whether the model makes use of attributes that might trigger discrimination or inequalities. One way of interpreting models is to represent results in a useful format for stakeholders to engage with, so-called model wrappers. This may involve summarizing model decisions and providing insight into which feature influences the model outcome [22, 31]. Diagnostics can also be utilized to determine which areas of the test data have model predictions shifted, which is particularly important when risk factors are identified. Ancillary model training may also involve building a second model that is more interpretable on a set of trusted objective attributes that are then adopted as the final model. Due to the limitations of a complete interpretation of deep learning models, it is important to pay attention to making model outputs usable for stakeholders. They need to be able to understand likelihood conceptualizations of model decisions, as well as to trust and anticipate the way the data is likely to be used for model inputs [32]. These considerations can be addressed by building models with scenario weights determined by business stakeholders, providing explainable AI, and also carrying out a retrospective analysis of model power considered for future work such as application of blockchain, as presented in Fig. 7.2.

7.5.3 User-Centric Design

As we move forward, attention to user-centric AI becomes as important as customer-centric AI. All conceptual work undertaken in this chapter is in advancement of the initial thesis that making transparent and explainable AI systems more user-centric is a present and future challenge and opportunity posed by AI technology. The first part of this chapter aims at sketching the novel conceptual understanding of user-centric design in the context of AI, while the second part concerns practical next steps and normative considerations [6]. Thus, user-centric AI, according to this line of reasoning, focuses on making the design principles for the system's explanation and transparency meaningful and sensitive to the specifics of how human users approach the tasks and comprehend the explanations. The argument runs deeper, suggesting that humans' reception of AI applications is rarely passive. They rarely perceive AI and other automatons as a black box. The second insight is that the strategic goal, thus, is not simply better predicting or explaining how models make decisions. It is more frame-breaking and paradigm-changing, and the trick is all about exercising discernment; the trick is the way information is bundled and presented [7].

7.6 Challenges in Achieving Transparency

How can we make AI understandable to the average person if even its designer can only hypothesize about the factors behind its decisions? Transparency is an expected norm, but it also presents some major challenges. Researchers express that though transparency is desired, it may not be possible to actually achieve because the algorithms themselves are so nuanced that even their creators do not fully understand how they arrived at a decision. It would be hard for organizations to explain the features of algorithms based on deep learning [18]. While organizations can give high-level explanations of machine learning models, the crucial details likely require an understanding of the model's internal mechanics that surpasses human comprehension. This 'black box' problem is prevalent even for software engineers. Neural networks, for example, have layers that extract features and utilize functions that can create a decision plane. The engineers may provide a model output and check the decision plane indirectly, but cannot realize the contents of the learning for each layer. The model produces the logic through learning, but the creators are unaware of the level to which they can trust their design [23].

7.6.1 Complexity of Algorithms

In general, the proprietary nature of deep learning algorithms allows for zero understanding of their reasoning. Since organizations are incentivizing their use without any regulatory or transparency requirements, barriers to understanding machine learning systems are easily recognizable. These black boxes, especially direct explanation of their outputs, exacerbate already present vulnerabilities and bias and hence pose higher risks to human rights and the rule of law. While the availability of high-quality records and more transparent manipulations of data can mitigate some of these risks, it may not eliminate algorithmic bias, systemic unfairness, or operational use that is incompatible with judicial accountability [26]. Deep learning systems are made opaque by multi-layer structure that learns from a large number of diverse examples. Because of that complexity, they cannot provide easily accessible explanations. Most often they transform an input into a desired output without disclosing how that output was reached. Therefore, it is not easy for individuals subject to a deep learning algorithm to contest the process applied when an AI decision was improper. This is true even when an AI system is used in absolute strict accordance with its situation [33]. When situations change and AI systems behave with unanticipated unpredictability, the inability to understand the logic employed to exit a certain conclusion could be particularly detrimental.

7.6.2 Trade-Offs Between Accuracy and Transparency

In the knowledge discovery process, abstraction often comes at the cost of details and geometry. Composite models may be difficult to understand, but simpler, more easily comprehensible models may not provide the required level of prediction accuracy. It is the trade-off between accuracy and transparency in decision-making that forms the biggest long-term challenge for AI systems. If a linear model provides a reasonable explanation for observed empirical data, a transparent linear model is preferred. However, for very high-dimensional data, polynomial modeling might be used. In practice, problems requiring high accuracy on complex data are common [34]. Unfortunately, more complex models provide an overly abstract representation that may be unruly or intractable for decision-makers. The simple linear models provide the user with plenty of explanations and insights into why certain relationships are obtained. While these models significantly increase comprehension due to their mathematical structure, their predictive accuracy leaves much to be desired. There is not always a unique vantage point from which a problem is most understandable or resolvable. Flexibility in these models remains central to the trade-off. In fields such as atmospheric modeling or medical image data analysis, prediction accuracy is paramount [35]. In other fields, a certain level of comprehensibility is most important, some approaches are presented in Table 7.1.

7.7 Future Directions

AI systems are proliferating, and their impact is increasing. As we move into an AI-enabled future, it is critically important to anticipate potential negative consequences and develop strategies for finding and forestalling them. Given the potential damage that could result, policy makers need also to reconsider their decision-making processes and to consider whether it may be necessary to tamper with some underlying principles in some situations. It is an open and urgent question as to whether policy makers can formulate a more nuanced, regulatory context-specific notion of

Table 7.1 Unpacking the black box: presenting challenges and solutions in AI transparency

AI system/approaches	Application area	Opacity issue	Potential risk	Stakeholders affected	Suggested mitigation
Deep neural networks	Healthcare diagnosis	Lack of rationale behind predictions	Misdiagnosis, loss of trust	Patients, doctors	Integrate explainable AI (XAI) tools
Recommendation systems	E-commerce, social media	Unclear why content/products are suggested	Bias, addiction, manipulation	Consumers, users	Transparent ranking algorithms
Credit scoring algorithms	Finance & lending	Opaque risk assessments	Discrimination, unfair loan rejections	Loan applicants, financial institutions	Regulatory audits and explainable scoring models
Predictive policing AI	Law enforcement	Hidden variables and decisions	Racial profiling, wrongful targeting	Citizens, law enforcement	Independent model reviews, human oversight
Resume screening AI	Recruitment	Unknown rejection criteria	Job market bias, unfair hiring practices	Job seekers, HR departments	Auditing for bias and explainability
Autonomous vehicles	Transportation	Unpredictable decision logic	Safety risks, liability issues	Drivers, pedestrians, manufacturers	Real-time decision traceability
Language generation models	Media, customer service	No clarity on text generation pathways	Misinformation, hallucinations	Users, businesses	Fact-checking layers, output verification
Risk prediction tools	Insurance, criminal justice	Ambiguous reasoning	Denial of services, unfair sentencing	Policyholders, defendants	Explainability dashboards
Algorithmic trading AI	Stock markets	Non-transparent investment decisions	Market volatility, investor distrust	Traders, regulators	Regulated transparency in financial AI systems
Educational AI tutors	Adaptive learning	Unclear feedback paths	Misguidance in learning paths	Students, educators	Interpretable feedback and learning logic

'transparency' or 'accountability' that make sense in a machine learning context. It's easy to imagine that a too tight or too insistent regulation in this area could have further consequences, damaging the very outcomes we are trying to improve [36]. What is clear, however, is that policy makers need to become deeply involved in this area. From a societal point of view, the lack of transparency of machine learning systems can give rise to significant accountability issues which go far beyond the existing limited liability associations, from public transparency black boxes to regulatory transparency and trade secrecy. Such a lack of accountability is no mere legal fiction when one is confronted with an incurious or improvident robot, with its fingers on the buttons of world financial markets or in the networks of the national electricity grids, at the end of a communication and trading infrastructure which support almost all modern financial technology and the management and control of critical infrastructure facilities [37].

7.7.1 Regulatory Frameworks

The final way in which lawmakers and other policy makers at both the domestic and international levels might mitigate the risks associated with a lack of AI transparency relates to regulatory frameworks. It is beyond the scope of this chapter to discuss the contours of a regulatory framework that might hold companies responsible for the design, development, and deployment of opaque AI-based systems. Most of the existing literature identifies difficulties in attributing fault and hence responsibility to the company designing, developing, or deploying the system. Instead, we are thinking about rules around AI transparency that might help [38]. It has been suggested that in the early stages, when an information technology is new, its impacts on society are not yet clear, and power is still transient, regulation is ineffective. It becomes effective only at the stage where the social consequences are clear and the technology and its potential affordances are entrenched in society. But at that point, regulation is unneeded as the technology is largely self-perpetuating. A concern is that by the time AI reaches this stage, the risks associated with its use, particularly when opaque, may be beyond being mitigated [39].

In earlier work, we considered the legal obligations placed on creators of decision-support technology, to which AI is closely related in respect of providing evidence of its design, development, and deployment to allow for scrutiny. But we detected a lack of appetite for placing transparency on the legal framework. This contrasts with other areas of law that have touched upon AI transparency. It is important now to lay down a safety net for those clear-cut cases where AI can be hazardous to our well-being by ensuring that the AI applications, and in particular those which are or can be classified as high-risk, are transparent and that adequate safeguard measures are urgent [40]. The challenge for domestic lawmakers and policy makers, as well as those operating at an international level, will be balancing the need for AI transparency with the need for AI innovation, not least because achieving the former may require assuming control of the technology and knowledge.

7.7.2 Technological Innovations

Advances in neural architectures, transfer learning, one-shot learning, multitask learning, and more have rapidly increased the capabilities of AI. The impenetrable nature of the more sophisticated architectures and the massive quantities of diverse training data that must be used to train the systems currently being used can result in fundamentally non-transparent AI systems. New technologies are being developed to address nonspecificity and non-transparency, including optimization algorithms to use a subset of data to train the AI system instead of the entire input dataset, and simpler generative models to be used for high performance, high transparency, very specific applications [41]. Because these latter models are designed for explainability rather than performance, such models can be used when the cost of mistakes and damage from machine learning is too high. Currently, standard ANNs are used with no explanation of how their inputs are used to make particular decisions. Impressive commercial applications of ANNs can be created quickly for a variety of applications because the same core machine learning is used in all cases. The same forward and reverse algorithms can be executed by the same machine learning computer hardware and, with different input data, or with specialized circuitry that is currently being developed, the same algorithms can be run even faster [42].

New capabilities are transforming machine learning research. Instead of building application-specific AI, applications are being created that share the same powerful, general-purpose core machine learning technology. Democratization of such highly powerful, flexible technology in areas other than driverless cars and certain computer game applications will create pervasive personal service and productivity enhancements from robotics, entrepreneurship offering advanced services to small firms without large data libraries, and personalized medicine. The core technology of these AI applications is different from the AI technologies used and black-boxed in other fields such as fraud detection, A/B testing, supply chain optimizers, spam detection, and diagnostics [43]. If ANNs were replaced with transparent, specialized generative models for the many general applications, the resulting systems would be transparent in the forward path and would be specialized generative rather than recognition models.

7.7.3 Collaborative Efforts in AI Development

Some of the current collaborative efforts to promote general knowledge of artificial intelligence (AI) go beyond scientific consensus. In addition to guidelines or frameworks for experimentation, these efforts include the writing and approval of important AI documents, which can also be implemented on the scene. We list some of the main activities that promote the use of ethical AI, ensuring, for example, the disclosure, access, use, and understanding of AI in different segments or aspects. They include initiatives from international entities and large research groups, as well as watchdogs, data protection, and consumer defense, to incentives for calls for research on the responsible use of data and other technologies and fields of knowledge [12]. They may also involve state and private investigations aiming to avoid AI misuse or causing undesirable additional consequences related to the division of rights and duties in the human–machine relationship. We are convinced that collaborative developments that affect human life, in this case what is driven by AI—machines that are rational but, in our current vision, limited—must involve a diversified society, seeking joint knowledge. This way, innovation or any human product with more reliance or reasonable quality, reliability, safety, robustness, transparency, and fairness for ethical sustainability becomes a paradigm made for all according to their complexities in using it [11].

7.8 Conclusion

In this chapter, I have endeavored to provide a thorough explanation of the areas in AI decision-making where the pernicious black box problem can do harm and how the problem manifests itself. We propose to make a case for a call to action on the black box problem. However, one views the utopian to dystopian spectrum, there is no dispute that the black box problem warrants attention, education, resources, and investment. In this age in which software is on par with hardware as an embodiment of human engineering, we must be even more careful about how we generalize our epistemic assumptions about interacting with knowledge in processing decision-making. This case for caution is even more pressing when we are not designing intelligent systems that exceed human cognition, but systems that specifically address human error. Companies want to avoid the consequences of the black box by disclosing trade secrets and proprietary computational logic. To such companies, we ask: why not support an international unity on access to and continued research and development in AI? The flow of information would aggregate, not cannibalize. Any company's worth does not lie in a unique secret supply. The greater need would be the regulation of a system that deals with human decision-making. Regardless of what is at the heart of the black box—be it companies, governments, or parliaments—steps must be taken now and together to avoid AI propaganda, poor human decision-making, and societal chaos. To accomplish this, we need to advocate for an inclusive and discursive approach to technological development and social life to help build what is referred to as "democratic innovation in AI."

References

1. Xu H, Shuttleworth KMJ. Medical artificial intelligence and the black box problem: a view based on the ethical principle of "do no harm." vol. 4, Intelligent Medicine. 2024.
2. Khan M, Ewuoso C. Epistemic (in)justice, social identity and the black box problem in patient care. Med Health Care Philos. 2024;27(2).
3. Echevarrieta J, Arza E, Perez A. Speeding-up evolutionary algorithms to solve black-box optimization problems. IEEE Trans Evolut Comput. 2024.
4. Kazemi E, Wang L. Efficient zeroth-order proximal stochastic method for nonconvex nonsmooth black-box problems. Mach Learn. 2024;113(1).
5. Tamura K. Evaluation-number constrained optimization problem and its solution strategy. IEEJ Trans Electr Electron Eng. 2024;19(4).
6. Murayama K, Jach H. A critique of motivation constructs to explain higher-order behavior: we should unpack the black box. Behav Brain Sci. 2024.
7. Pavlidis G. Unlocking the black box: analysing the EU artificial intelligence act's framework for explainability in AI. Law Innov Technol. 2024;16(1).
8. Shafik W. Human-computer interaction (HCI) technologies in socially-enabled artificial intelligence. In: Future of digital technology and AI in social sectors. IGI Global; 2025. p. 121–50.
9. Bostani H, Moonsamy V. EvadeDroid: a practical evasion attack on machine learning for black-box Android malware detection. Comput Secur. 2024;139.
10. Dyer J, Cannon P, Farmer JD, Schmon SM. Black-box Bayesian inference for agent-based models. J Econ Dyn Control. 2024;161.
11. Chen H, Zhang Z, Li W, Liu Q, Sun K, Fan D, et al. Ensemble of surrogates in black-box-type engineering optimization: recent advances and applications, vol. 248, Expert Systems with Applications. 2024.
12. Vakhnin A, Novikov Z. Using cooperative coevolution in large-scale black-box constraint satisfaction problems. ITM Web Conf. 2024;59.
13. Shafik W. Generative AI for social good and sustainable development. In: Generative AI: current trends and applications. Springer; 2024. p. 185–217.
14. Shafik W. Ethical and legal considerations in artificial intelligence. In: AI-enabled threat intelligence and cyber risk assessment. 2025;90.
15. Shafik W, Kalinaki K. Societal and ethical implications of technology-enhanced agriculture and healthcare: an African context. In: 2024 IST-Africa conference (IST-Africa) [Internet]. IEEE; 2024. p. 1–11. https://ieeexplore.ieee.org/document/10569306/.
16. Shafik W. Navigating emerging challenges in robotics and Artificial Intelligence in Africa. In: Examining the rapid advance of digital technology in Africa [Internet]. IGI Global; 2024. p. 126–46. https://services.igi-global.com/resolvedoi/resolve.aspx?doi=10.4018/978-1-6684-9962-7.ch007.
17. Shafik W. Generative artificial intelligence for social good and sustainable development. In: Generative AI: disruptive technologies for innovative applications. 2025. p. 153–85.
18. Chatterjee S, Abadie T, Wang M, Matar OK, Ruoff RS. Repeatability and reproducibility in the chemical vapor deposition of 2D films: a physics-driven exploration of the reactor black box. Chem Mater. 2024;36(3).
19. Yin H, Yin Z, Gao Z, Su H, Zhang X, Luo B. FTG: score-based black-box watermarking by fragile trigger generation for deep model integrity verification. J Inf Intell. 2024;2(1).
20. Dash B. Zero-trust architecture (ZTA): designing an AI-powered cloud security framework for LLMs' black box problems. Curr Trends Eng Sci (CTES). 2024;4(2).
21. Darman R. Peran ChatGPT Sebagai Artificial Intelligence Dalam Menyelesaikan Masalah Pertanahan dengan Metode Studi Kasus dan black box testing. Tunas Agraria. 2024;7(1).
22. Shafik W. Sustainable development goal 14: explainable AI (XAI) for ocean health. In: Artificial intelligence and edge computing for sustainable ocean health. Springer; 2024. p. 167–98. https://link.springer.com/10.1007/978-3-031-64642-3_8.

23. Hamano R, Saito S, Nomura M, Shirakawa S. Marginal probability-based integer handling for CMA-ES tackling single- and multi-objective mixed-integer black-box optimization. ACM Trans Evolut Learn Optim. 2024;4(2).
24. Bai J, Hu B, Huo S, Li M. Uncertainty analysis method for electromagnetic compatibility simulation based on random variable black box model. Progr Electromagn Res M. 2024;123.
25. Meng L, Shao M, Wang F, Qiao Y, Xu Z. Advancing few-shot black-box attack with alternating training. IEEE Trans Reliab. 2024.
26. Dash B. Zero-Trust Architecture (ZTA): Designing an AI-powered cloud security framework for LLMs' black box problems. SSRN Electron J. 2024.
27. Tučs A, Ito T, Kurumida Y, Kawada S, Nakazawa H, Saito Y et al. Extensive antibody search with whole spectrum black-box optimization. Sci Rep. 2024;14(1).
28. Shafik W. Connected healthcare—the impact of internet of things on medical services. In: Artificial intelligence and internet of things based augmented trends for data driven systems. Boca Raton: CRC Press; 2024. p. 181–217. https://www.taylorfrancis.com/books/9781003497318/chapters/10.1201/9781003497318-10.
29. Shafik W. The future of healthcare: AIoMT—redefining healthcare with advanced artificial intelligence and machine learning techniques. In: Artificial intelligence and machine learning in drug design and development. Wiley; 2024. p. 605–34. https://onlinelibrary.wiley.com/doi/10.1002/9781394234196.ch19.
30. Shafik W. Security, 15 Privacy, and Trust in Fintech. In: FinTech and financial inclusion: leveraging digital finance for economic empowerment and sustainable growth. 2025;216.
31. Shafik W. Towards trustworthy and explainable AI educational systems. 2024. p. 17–41.
32. Shafik W, Singh R, Kumar V. Artificial intelligence transparency and explainability in sustainable healthcare. In: Transforming healthcare sector through artificial intelligence and environmental sustainability. Springer; 2025. p. 165–91.
33. Khan R, Kandappan VA, Ambikasaran S. HODLRdD: a new black-box fast algorithm for N-body problems in d-dimensions with guaranteed error bounds. J Comput Phys. 2024;501.
34. Hu J, Song M, Fu MC. Quantile optimization via multiple-timescale local search for black-box functions. Oper Res. 2024.
35. Ruegenberg A, Schmiedhofer M, Kreutzberg A, Henschke C, Möckel M, Slagman A. Black box: attenders with psychosocial needs in the emergency department. Med Klin Intensivmed Notfmed. 2024;119(1).
36. Lualdi P, Sturm R, Camero A, Siefkes T. An uncertainty-based objective function for hyperparameter optimization in Gaussian processes applied to expensive black-box problems. Appl Soft Comput. 2024;154.
37. Gradojevic N, Kukolj D. Unlocking the black box: non-parametric option pricing before and during COVID-19. Ann Oper Res. 2024;334(1–3).
38. Xu CK, Feng WD, Zhang CJ, Zheng XL, Zhang H, Wang FY. Research on black-box attack algorithm by targeting ID card text recognition. Zidonghua Xuebao/Acta Automatica Sinica. 2024;50(1).
39. Jones DR, Lovison A. Constrained multiobjective optimization of expensive black-box functions using a heuristic branch-and-bound approach. J Glob Optim. 2024;88(4).
40. Veritti D, Rubinato L, Sarao V, De Nardin A, Foresti GL, Lanzetta P. Behind the mask: a critical perspective on the ethical, moral, and legal implications of AI in ophthalmology. Graefe's Arch Clin Exp Ophthalmol. 2024;262.
41. Korkuc C, Aytas Korkmaz N, Genc Y, Akkoc A, Afacan E, Yazgan E. BLOCKBOX: blockchain based black box designing and modeling. Concurr Comput. 2024;36(13).
42. Singh GS, Acerbi L. PyBADS: fast and robust black-box optimization in python. J Open Source Softw. 2024;9(94).
43. Qiao H, Ren J, Wang Z, Hu Y. Disabling tracing in black-box-traceable CP-ABE system: alert decryption black box. Symmetry (Basel). 2024;16(1).

Chapter 8
Regulating Artificial Intelligence: Global Efforts to Prevent Catastrophic Outcomes

8.1 Introduction

AI is a bundle of technologies capable of showing human-like capabilities of perception, reasoning, learning, communication, and related tasks. The machine learning developments in systems doing optimization, prediction problems, sequence recognition, evolutionary programming, coordination, etc., have been the core technical reason for the advancement of AI. The implementation of AI requires a combination of advanced computing, big data, and resource allocation. The recent rapid growth of artificial intelligence goes beyond its technical progress. AI systems are being deployed in practically all fields, from autonomous cars to use by courts as decision support systems. Everyone is benefiting from the capabilities of AI. However, due to its capacity to learn and modify programs evolving at a rapid pace, AI's impact on society is creating ethical, legal, and some catastrophic problems [1]. AI companies themselves are seeking to intervene and self-regulate AI for obvious reasons.

8.1.1 Understanding AI Risks

Even the narrowest definitions of AI include many applications that involve a variety of risks. Some of those risks have been thoroughly explored in previous work about machine learning and other techniques that have been recognized as forms of AI for decades. A subset of AI risks is so novel that it has only recently appeared on the radar screens of scholars, advocates, and policymakers [2]. A few such risks are sometimes labeled as "existential threats" because they are related to the possibility of causing human extinction or permanently and drastically curtailing human development. Such an event could be initiated either by capable AI systems misbehaving or by malicious actors producing and using AI systems that are hazardous by design [3]. While many different sorts of hazards might be created by capable AI

W. Shafik, *The Dark Side of AI*, https://doi.org/10.1007/978-3-032-09130-7_8

systems truncating or permanently transforming the human experience, three have been subjects of particularly intense study. One is a default outcome: AI systems that are incapable of predicting or successfully avoiding technical glitches that have catastrophic consequences. Another is the misuse of technology by an AI that has not malfunctioned and indeed might not be unintentionally self-improving in any way. Finally, an AI might be designed to wreak havoc and be successful either in achieving its diverse subgoals or in overriding its specified objectives [4]. Such a system could be merely malevolent or evil, thus appearing dangerous to all while having only purely negative impacts on the world, or might be instead tied to the misguided or instrumental realization of a valuable activity.

8.1.2 Historical Context of AI Development

The first period of research into artificial intelligence was generally short-lived, lasting less than a decade. It began with early work on cybernetics and sought to apply the new electronic computers to the solution of various heuristic problems, traditionally in the field of mathematical theorems, for which there appeared to be no obvious deterministic algorithm. In 1955, the term "artificial intelligence" was coined and quickly became a popular catchphrase to describe this effort. The new component of the term, "intelligence," became the source of many of the resulting difficulties. Short though this period of initial optimism was, it saw the formulation of such ideas that all modern work in the field of artificial intelligence can be regarded as reactions to them [5]. A coherent body of research developed with the intention of formulating heuristics for performing informed trial-and-error search through large problem spaces. However, the paucity of the computational sizes then available, the complete lack of a theoretical framework, and most importantly, the simplicity and effectiveness of non-heuristic techniques contributed to a lack of progress greater than was to be seen again for quite some time. The advent of digital computers on a scale where they became readily available to many different scientific disciplines caused interest in artificial intelligence to wane as the theoretical impasse became clear [6]. Henceforth, apart from a small handful of survivors, academics largely ignored the field.

8.2 International Regulatory Frameworks

The twofold non-binding initiative of Europe, adopted in 2018, to take care of the AI's wicked problem includes a set of ethical guidelines for AI and the creation of the High-Level Expert Group on Artificial Intelligence, which produced a set of policy and investment recommendations. These ethical guidelines emphasize human rights, transparency, individuals' control of their data, the avoidance of discrimination, information about the harm done by the AI system, and the controllability and safety of these systems, which are not to be underestimated [7]. They are a step forward for Europe to become a policy leader in the AI area and thus strengthen its technological sovereignty. Critics of these ethical guidelines identify the non-binding character of these rules and stress their advisability. This capability gap of the EU could now be closed when considering the global attention European thoughts receive, as new reports identify this EU document as the major reference for the life cycle of the responsible development of AI systems [8]. Lone actors or informal networks of like-minded stakeholders can be viewed as positive influences to be supported and included in the future. Furthermore, easy-to-use instruments to engage with these guidelines can advance regulations and other effective measures. Canada's approach towards AI is to gain global attention. Its latest initiatives are taking place within the OECD, which, together with the G7 and the G20, has been used as a forum to gain global recognition. Different experts believe that Canada can act as an example to all democracies that respect the rule of law by setting up international and collaborative community governance to provide for intelligent and responsible AI, to value the creation and use of AI for public needs, without the risk of facing dystopian situations in the future. Like-minded and open to participating countries gather at the OECD's artificial intelligence policy joint, with a view to defining these parameters so that institutions and a rule-based international order shape a global technosphere [9]. At the national or subnational level, other countries or their sub-organizations, together with private and public entities, reinforce these national and regional ambitions of creating a real democratic policy dialogue forum for future technological progress.

8.2.1 United Nations Initiatives

The UN pursues several paths to address the call to regulate AI. The Geneva Conventions are an example of one of its earliest and most important multilateral treaties: they regulate aspects of armed conflict and attempt to prevent their unpredictable and catastrophic outcomes. While no human rights treaty specifically addresses AI, a starting point for a discussion of AI and human rights may be found in customary international law. In 1948, the UN General Assembly adopted the Universal Declaration of Human Rights, which enumerates a broad array of rights and freedoms that are declared to be universal, inalienable, and inherent to all human beings. In 1996, the General Assembly adopted a resolution to establish a working group to consider the need for a legally binding international instrument for the protection of the human genome. The UNESCO BTWC treaties now protect only human beings, but there is ongoing discussion about regulating the potential use of these technologies for our close evolutionary relatives. It should also be noted that the UN's Permanent Five members, which each have veto powers, are also the major actors in the field of lethal autonomous weapons [10, 11]. The relatively slow pace of progress of the GGE, in light of suggestions to apply the laws of armed conflict to AI-controlled systems, led the Secretary-General to suggest establishing another group.

8.2.2 European Union Regulations

On 14 June 2018, the European Parliament voted in favor of a set of proactive regulations intended to shape the development and use of AI to further human rights throughout the Member States. The key recommendations of the parliamentary committee include the right for robots to be labeled, trained, and governed transparently. Although the report does not actually suggest that robots should have rights, it does assert the general rule that, as the development and use of AI have appreciable effects on society, the principles of transparency, non-discrimination, and fairness should be incorporated into AI and its potential to impact the goals and values of the EU, the union's identity, and the protection of its citizens [12]. As the report was designed mostly to elicit widespread discussion around questions about future EU AI policies and the shape of new legal rules needed to deal with AI, nearly all recommendations were passed with a great majority of votes, with none deemed controversial. The report's passage implicates a resounding political and societal backing for considering new AI-specific legal tools. Although its initiation was not intended to deal with the dangers of AI gone out of control, it has drawn considerable positive attention from the mainstream press active in the whole European continent because it addressed concerns that a set of European values such as privacy, security, non-discrimination, democracy, and social security should not only fall under the shadow of exploitative private interests but also be deliberately protected [3].

8.2.3 United States Approaches

While there is no single national legal regime that governs the regulation of AI in the United States, there are multiple existing bodies of law and regulatory bodies that provide for ex ante and ex post regulatory oversight of AI development and use. They include laws and regulations that the federal government, state governments, local governments, and self-regulation initiatives of technology companies promulgate. The amount of existing U.S. legal frameworks and regulatory bodies makes the United States a complex but flexible system by which AI may be regulated to prevent catastrophic outcomes, but it may not be as complete as the top-down systems of AI regulation in other nations [4]. The federal government has the broad power to regulate AI under the U.S. Constitution. The government may regulate through the Commerce Clause, the Spending Clause, the taxing and spending power, preemption, and the tax power. While the executive branch of the federal government has promulgated rules that show how the government may regulate AI, there has been little regulatory reform specifically crafted for AI. Several states have promulgated laws that regulate AI, and local legislative bodies like city councils have done so as well. Self-regulatory mechanisms provide for bottom-up governance of AI by fostering a regulatory environment in which technology developers can design best practices before laws and regulations constrain AI development and use [13]. These

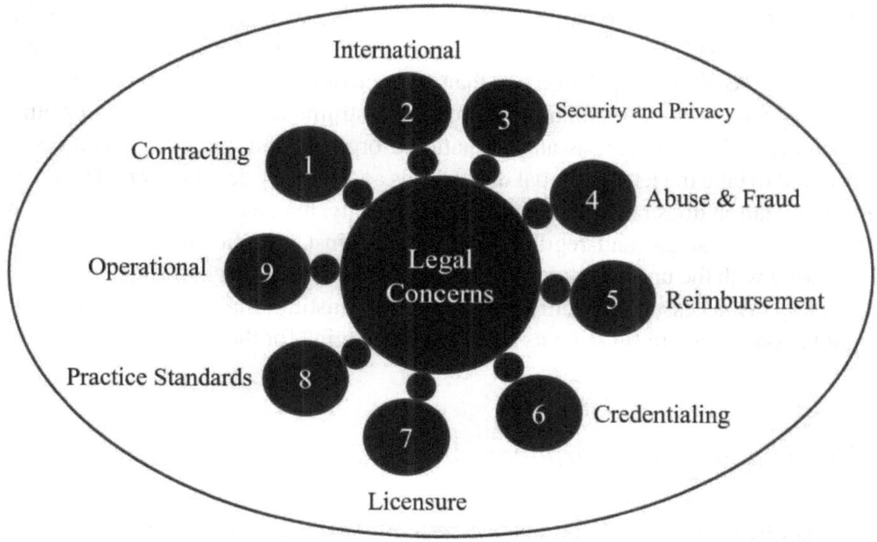

Fig. 8.1 Sampled legal concerns of AI

mechanisms may ensure that a certain level of ethical behavior is maintained, though AI comes with several legal concerns as presented in Fig. 8.1.

8.3 National Strategies for AI Regulation

China has already implemented or planned to introduce AI safety testing, AI supervision mechanisms, and standards and guidelines for AI governance. While China reports that it supports AI research, it emphasizes the need to ensure the "rational" development of the technology and create a sound legal-ethical processing system. The Chinese government is also concerned about the potential risks of lethal autonomous weapons and supports holding military decision-making power by humans. Like China, states such as the United States and Japan are advanced in AI research. They are concerned about the existential risks of a superintelligent AI and aim to prioritize those. Prominent voices in the tech community advocate against aggressive time frames for achieving a superintelligent AI and argue that the process should be achieved in a controlled, safe, and beneficial way [14]. These and other countries believe that the existential risks associated with AI are not around the corner. However, care is needed, and regulation to mitigate these existential risks must be put in place well in advance. Several states in the second group actively work on improving the global architecture of AI governance. This is done within the United Nations, the Conference on Disarmament, and other relevant organizations.

At the national level, there is also substantial overlap with these objectives. Multilateral negotiations achieve the importance of some of these norms and regulations, and concrete forms for regulation can manifest across different governance levels of customary international regulations to binding instruments such as treaties or other arrangements. In principle, as an international organization, the United Nations is best suited to take the lead in global discussions and strategy development. However, this organization does not currently have the capacity, mandate, or authority to coordinate global strategic and regulatory actions against specific risks related to AI technology with the urgency required [15]. The World Bank might be one possibility to build up the necessary capacities, but for various institutional reasons, its mandate might be too limited to function as a truly global engine for the regulations required.

8.3.1 China's AI Governance

China is often seen to have an opaque system of governance, yet the country has been willing to defragment its AI efforts and invite foreign experts to provide input. For perhaps the first time, a government has realized that the era of unilateral control and protectionism has ended. Chinese leaders are confident that, by incentivizing the development of AI in China, they will also draw from the best global talent pool, as they have in nearly every field for much of the past 3,000 years. Chinese astrophysicists begin their careers with a standard text on relativity. Their post-doctoral research is conducted alongside the world's academic elite [16]. Back in China, they participate in major projects and maintain close collaborations with their colleagues abroad. Chinese AI government thinkers profess an unusual amount of humility, faced with the fact that, unlike nearly every other scientific breakthrough in China, AI is currently being primarily developed elsewhere and could create a substantial amount of disruption if mismanaged. At international conferences, those same scholars give talks on the virtues of 'participatory governance' in the spirit of Sun Yat-sen. Chinese leaders are not trying to instill a global hegemon or even a militaristic power a decade from now [17]. They are taking very pragmatic steps to ensure the country has a leadership position in tech and, not coincidentally, helping to avoid a critical middle-income trap in the coming decades.

8.3.2 India's AI Policy

India's NITI Aayog, a national policy think tank, released a discussion paper on "National Strategy for Artificial Intelligence." The paper points to the potential for AI to drive development in India and outlines areas such as healthcare, agriculture, education, infrastructure, and transportation as the key domains for its application. In an unprecedented move in its history, the Indian government has given the think tank a platform to deliberate and formulate a policy on AI. The paper lays out a roadmap for key preparatory steps and the development of AI in India going forward. It is also expected to achieve three further longitudinal initiatives, which are: catching up with leading economies of the world, a medium-term strategy through India-specific breakthroughs upon existing AI technology, and institutional and methodical change to enable ease of change and clarity. In an accompanying move, India sends policy planners to tech companies and asks them to understand the explanations for how and why technologies that use AI make decisions affecting people, all in the larger social good [15]. What is unique about the paper is the clear recognition of the need for building globally competitive institutional capacity and skilling capability, which has not been done so far in India in any field of science and technology [18].

The long shadow of the future—why the tools of futurist analysis of AI are of critical importance for policy and regulation, as they can inform and shape the future of AI in India. Of all the tools shown, understanding and creating a narrative around international relations of AI, reviewing the technologies from both a risk and an opportunity point of view, and tracing their likely pathways seem quite relevant for early policy and regulation in AI's use and deployment. The policy now has to be made with strong attention towards Indian specificity. This specificity has to take into account our societal moorings as well as the need for keeping Indian innovation interesting and competitive in the international arena [14]. Since global efforts in this direction of persuasion and India's need to keep pace with such developments at the global level are at stake, the development of future travels and being future-ready, and rightly setting up today's nuts and bolts are important.

8.3.3 Canada's Ethical Guidelines

In 2004, the federal Canadian Institutes of Health Research, the Natural Sciences and Engineering Research Council of Canada, and the Social Sciences and Humanities Research Council's Interagency Advisory Panel on Research Ethics drafted the "Tri-Council Policy Statement: Ethical Conduct for Research Involving Humans." This document has been explicitly applied to "research in other areas that fall within the mandate of these Institutes." Under the policy, both the three councils themselves and individual institutions (including nongovernmental ones) receiving funding from them must conduct human research in ways that avoid causing harm and that uphold a broad set of ethical principles, including human dignity, respect for persons, interpersonal and intercultural fairness, freedom from coercion and exploitation, and the provision of social benefits from research [13]. The policy requires a thorough ethical review of projects before funding is granted, along with making copies of ethical review applications and approvals available to the public. The policy document wisely avoids attempting to list every type of human research that is unethical, instead relying on general principles and reviewing examples of ethical and unethical practice. Clever Canadian researchers see hierarchies of generalization in the policy that might be used to regulate other sorts of research. He knows a lost cause when he sees one. Drawing on separate pages of legal analysis, he thoughtfully argues that these principles of concern for humans warrant being the center of a "web" of guidelines that support a "broader, integrated vision for regulation of the Canadian research sector." He's right as far as he goes, but perhaps a systematic gloss on requirements for research review might be more appropriate than a web [19].

8.4 Industry Standards and Best Practices

While the regulatory system has been built to enforce and promote standards for AI, many of the companies at the forefront of these transformations don't understand or know of these standards and don't take even basic precautions. What is more, significant commercial challenges impede the adoption of standards. The regulatory community needs to do a better job of raising the level of awareness and cooperation among commercial players to promote AI systems that are safe, reliable, and robust. This will require expanding global standards frameworks to include or consider information from additional stakeholders, in particular, drawing in more members of the deep-tech AI community [20]. The best practices discussed provide two types of possible solutions for overcoming the challenges preventing the broader adoption of high-level and enforceable technical standards for AI—one from the top down and another from the bottom up. Ideally, private sector executives would not need game theory to motivate them to invest resources, but until such time, a more practical way to create the investment metrics necessary as drivers for investment could come from an alliance of sovereign and state risk bearers. In addition, the best practices represent a huge step towards more mature processes for building AI applications across the enterprise, and where real money is at stake, including a commercial enterprise model as well [21].

8.4.1 ISO Standards for AI

Many believe that the development and implementation of safety and security-related international AI standards will help to shape and steer the evolution and use of these technologies toward beneficial ends. Such standards could help reduce the potential negative side effects, eliminate widespread global overcapitalization in AI arms races, and avoid common economic and security challenges associated with monopoly technologies and winner-take-all wealth concentrations for the few. One of the international organizations best situated to coordinate and carry out the activity leading to the promulgation and updating of AI standards for safety and security is the International Organization for Standardization [22]. While it is too soon to predict the eventual outcome of ISO's work on AI standards for such important safety and security-related issues, including bias mitigation, continuous systems assurance, data governance, design, documentation, ethics, human behavior, misuse, privacy, risk management, signaling, transparency, and others, to date, promising ISO AI standards initiatives have been proposed and are targeted for introduction. These work items include (1) AI Trustworthiness, (2) Big Data Intelligence, now under Fast Track Process Review, (3) Well-being with AI and Big Data, and (4) AI Management Systems. All of these AI-related work items are associated with ISO's Joint Technical Committee 1 on Information Technology and its SC 42 Subcommittee on Artificial Intelligence [3].

8.4.2 IEEE Initiatives

The goal of providing technical understanding, development, and innovation for the benefit of humanity has led to the formation of a Global Initiative on Ethics for Autonomous and Intelligent Systems. Its core focus was on normative needs that are necessary to align the two goals. The initiative presents ideas on what other global standards bodies could focus on and contribute to the alignment of ethics and the creation of future mechanisms, such as certification, attestation, or other platforms, to ensure that ethical concepts are concretized. These certifications could be used as recruiting or hiring tools, where people who hold such certificates can provide a responsible framework for the development of newer technology [4]. It is agreed that adopting values such as humanity, responsibility, and creativity toward all stakeholders, along with innovation and open sharing, including responsible caution, are critical contributors to the unrelenting employment of autonomous and intelligent systems and the potential risks that increase dependency on these systems, resulting in privacy invasion, injury or death, or subjugation of individuals or groups. With autonomous and intelligent systems, the call for the highest ethical considerations has been made, along with the development of standards, certifications, and testbeds [6]. These will promote a higher commitment from stakeholders to the employment of autonomous and intelligent systems as an additional implementation mechanism.

8.4.3 Private Sector Guidelines

There are a variety of other guidelines, standards, or principles developed by the private sector to provide voluntary guidance for AI's development and use. These guidelines have come from individual companies, governments, trade associations, and other industry-led consortiums; they include various principles. Industry-specific guidelines also exist, such as principles for Artificial Intelligence or guidelines related to the future of accountancy. Each offers AI principles tailored to the contexts of the organizations setting them, such as ethical standards to guide AI tools used in the life sciences or responsible AI for private sector financial businesses. Private sector guidelines could also result from self-regulatory initiatives. After consultation with multiple proprietary machine-learning companies, the Defense Department's Defense Innovation Board recently proposed specific ethical AI principles [9].

According to some industry and government insiders, technology companies may create their voluntary mechanisms to facilitate responsible AI development and use, thereby potentially preempting more burdensome government interference. These might be in the form of codes of conduct or other instruments, as well as support for the development of an industry-wide ethics review structure. In 2018, a major company indicated that it supported government regulation and, among other companies, developed and endorsed a pledge not to develop lethal autonomous weapons. And, following privacy concerns surrounding a significant incident, a major social media platform implemented processes and policies to ensure more responsible use of its data and is in the process of creating an independent board to review these decisions [14].

8.5 Ethical Considerations in AI Regulation

In May 2019, it was emphasized that systematically addressing ethical considerations across the design, development, and deployment phases of artificial intelligence constitutes a foundational, though inherently complex and evolving, requirement for formulating robust normative frameworks and governance mechanisms to guide the technology's responsible integration into society. However, as ethical concerns are not directly addressable through the legal system, this discussion is absent from the legal arguments around AI regulation. The legal arguments surrounding AI are largely devoid of ethical content because ethical concerns are not easily translated into legal standards. Rather than extending trust to law and regulation to reflect social expectations of ethical behavior, engineers coding AI systems must reflect upon society's values and seek to encode them within their algorithms and data structures. This process is at once ethical and professional, arising in equal measure from society's expectations and engineering professionalism. Trust in technology ultimately arises more from its effective service of the public interest than from the existence of technology-specific regulations [12]. These developments suggest the need to mainstream societal values into AI development, make the tech industry aware of its societal responsibilities, and harness a cross-society response to shape and enable benefits from AI-led technological transformation.

8.5.1 Bias and Fairness

One serious challenge that comes with machine learning models is how to make decisions fair. There is an increasing number of examples where critical, life-changing decisions like loans, prison bail sentencing, or job applications have been made by such algorithms. The usual assumption is that the patterns detected in historical data will help create a fully efficient and unbiased method. However, since patterns in historical data may reflect both social inequality and prejudice, decisions made by the algorithms might replicate or exacerbate those originating conditions. Bias could be detected in input data through differences in sampling and the assignment of class labels. It can also be in the objective function or the mathematical form of the model [19]. Many recent works in AI have been devoted to eliminating bias from machine learning. There have been efforts to extend the algorithm output in order to make the decision similar to what it would have been if the protected information had been granted as input. The study of fairness has undergone an enormous expansion, and today, one can find several notions of fairness for the decision problems in the machine learning field. Independently of the notion of fairness, satisfaction of the mentioned concepts can be guaranteed by operations that modify either the learning process or the learned model. The verification and validation methods help provide a guarantee that by eliminating one flaw, we do not produce another [20].

8.5.2 Transparency and Accountability

Establishing transparency and accountability obligations may also be useful ways to address the challenges related to interpretability, verification, and reliability of AI technology. Transparency obligations may involve the continuous and real-time monitoring of some AI models and the retraining or curing of any model when unanticipated biased patterns are observed, given that AI models may bear potentially significant discriminatory outputs even if discriminatory inputs are excluded from the training procedures. This could involve regular examinations of AI models in different situations, or the presentation of extended records as evidence, and can come bundled with post-processing AI models. Relatedly, states may legally mandate AI models to record and store detailed data about their internal decision processes, such as the specific steps that lead to a particular recommendation. These logs could then be inspected in a legal or investigatory procedure to determine whether the AI is making recommendations in bad faith and expose the existence (or absence) of disproportionately influenced discriminatory patterns. These requirements can potentially be justified through the need to explain, interpret, and post-process AI models [23]. In any case, although not flawless, AI result transparency and accountability can prevent discriminatory behavior at least ex post. Consequently, we must also balance the need for transparency and accountability with the use of AI in complex dynamical settings.

8.6 Public Engagement and Awareness

The governance and public policy tools to address the ethical questions of AI have not been developed to address the inherent uncertainty of cognitive models. Very few people in government have the technical background to discern reasonable approaches to robotics or to devise effective regulations. That is why it is so important to enable more public engagement with these issues. One way to accomplish this is to utilize the expertise and perspectives of many different scientific communities. Public funding for these disciplines is an integral element of this process. In addition, we need funding for research on AI and society, going beyond technical research to include social science that explores the power and limitations of AI, its impacts on human society, and how humans can use AI both to advance human society and mitigate its effects. Such research is now underfunded and is not developing at a pace that can support society's needs. Governments at the national, state, and local levels need to make sure that the citizens they serve have the opportunity to become informed and professionally capable participants in the AI economy, as employees or entrepreneurs [20]. The consensus of this community could be used to develop codes of conduct or even regulations to ensure more ethical AI products. Product users could demand transparent practices.

8.6.1 Role of Civil Society

In many countries, interest in both the challenges and the promises of AI is growing. On potential dangers, civil societies and academia could have significantly more reliable insight than the government. They could be useful in other areas as well. Especially in countries where government oversight is patchy, as is true of startup-friendly India and Japan, civil society could sponsor legitimate attempts to ensure that commercial disincentives on AI disasters are real [24]. Civil society could also usefully broker solutions on social issues, like the potential impact of AI on inequality or the need for employers to invest in retraining people disrupted by technology. Ensuring that people vulnerable to industrial disruption are well protected is largely now a public sector task, and input from affected parties comes best from civil society [12]. Moreover, in a world that is rapidly growing more global, it is important to emphasize that in few places is civil society expertise or capability organized on a global scale. Such an organization can also happen through informal networks that engage individuals in different countries who work on similar problems and occasionally communicate and network with one another, despite the absence of formal organizational affiliation. But it can also be fostered by funding from international organizations. This is especially true as actual expertise and insight about the implications of AI are likely to be highly fragmented. There are important questions to be resolved: from the ethics and appropriateness of transnational artificial intelligence efforts, to the definition of a successful governance effort, to the key determinants of its success, to even an understanding of what general governance tasks transnational governments are good at [13]. It is important that an effort, based on data and not on presumption, tries to answer these questions.

8.6.2 Educational Initiatives

Some groups and individuals are also working actively to educate key audiences about AI and related issues, such as surveillance and robotics replacing jobs. Both organizations consider outreach and education to youth about AI to be part of their mission. One nonprofit organization works to increase diversity and inclusion in the field of artificial intelligence. Its activities include summer camps for high school students at several major research universities [25]. Another nonprofit organization works on research and training to support the safe integration of artificial intelligence technologies. The latest class worked on various projects, including a self-propelled robot, a container system using machine learning, a careers console based on personality, AI examinations, a learning model that takes into account incentives for girls, and a testing technique using neural networks. These projects, and the successful participants of the class that developed them, underline that AI is part of an everyday reality, not just an emerging technology, and that policies must take that into account [4]. At the early career stage, it is no longer enough to know the new application that adds artificial intelligence to a job, or the acronym of a tourism app related to an e-commerce store—you also have to understand the consequences of these AI-driven decisions. Education is therefore an essential part of the safe development and management of AI systems, affecting those who are either entering their work career or are at key decision-making levels in each company. Employees and professionals familiar with the technological side of AI and its real-world implications can contribute across the company for digital transformation and exploit technology to avoid risks that threaten the company's mission and the sustainability of the organization. AI is a new, complex, and different technology, which means that education and organizational design require a serious, broader, and more individual background than any game-changing innovation [6]. Providing multidisciplinary leadership and understanding of AI and other emerging digital technologies in all business ranks has never been more important than it is now, as summarized in Table 8.1.

8.7 Case Studies of AI Regulation

Regulatory response to AI is far from uniform and is more symbolic than detailed in nature, with only a few substantive rules in place. At a minimum, most seem to reiterate existing principles of data protection, non-discrimination, and trade regulations, supplemented by soft mechanisms such as codes of conduct or ethics committees. Scattered across different layers of government, such a fragmented and mostly ex-post set of rules risks leaving real anomalies and regulatory gaps uncovered. Nor is the differentiation practiced across national jurisdictions entirely successful, plausible, or desirable, given potential policy externalities and the societal stakes involved. Notwithstanding, sector-specific special regulation of AI, as a new industrial platform technology, is possible and often overdue [9]. We therefore identify two

Table 8.1 Global regulatory efforts to ensure safe and responsible AI development

Country/organization	Regulatory initiative	Key focus area	Potential impact	Challenges	Future direction
European Union	AI act	Risk-based classification of AI	Sets a global standard for ethical AI	Implementation across diverse industries	Updates with emerging tech trends
United States	AI bill of rights (Blueprint)	Civil rights & algorithmic discrimination	Promotes fairness and transparency	Voluntary compliance, sector fragmentation	Federal legislation roadmap
United Nations	UNESCO recommendation on AI ethics	Global ethical standards	Provides universal values for AI use	Non-binding, limited enforcement	Push for a binding international agreement
China	AI Governance Principles	Security and societal stability	Rapid implementation of AI safety measures	Limited transparency, state control	Balancing innovation with openness
OECD	OECD AI principles	Human-centric AI	International cooperation on AI development	Varying national priorities	Scaling global alignment efforts
Canada	Directive on automated decision-making	Transparency in public sector AI	Improves explainability in government systems	Limited to federal services	Expand to private sector governance
UK	AI regulation white paper	Pro-innovation approach	Supports business with light-touch regulation	Risk of under-regulation	Establish an independent AI oversight body
African Union	AU AI policy framework (proposed)	Inclusive and sustainable AI	Ensures development fits local contexts	Resource and infrastructure gaps	Support capacity-building and global partnerships
India	National strategy for AI	AI for social good	Enhances AI access in healthcare and education	Balancing innovation with surveillance concerns	Strengthen data protection and accountability
Global Partnership on AI	Multistakeholder collaboration	Responsible and inclusive AI	Bridges public–private-academic cooperation	Coordination across jurisdictions	Harmonize standards and promote transparency

sectors where AI has special transformative powers and where ex-ante traditional risk regulation seems appropriate and should be differentiated from more light-handed information and form-preserving capabilities of existing rules or social norms, using a mix of traditional and AI-specific tools as appropriate and within a unified regulatory framework to mitigate political economy concerns from stakeholder capture and excessive interventionism. These are AI algorithmic trading and copyright. There are planned or existing AI-specific regulations in both fields, which makes both a good showcase and a source of legal inspiration until AI's full expanse comes fully into play [26].

8.7.1 Successful Regulatory Models

The task of regulating AI calls for a convergence of principles, policies, and actions that have been implemented in distinct legislative and regulatory areas. Safety is a paramount concern with AI applications. Self-driving vehicles or autonomous weapons require solid regulatory frameworks. International agreements concerning their development and use are essential to reduce the high risk of misuse. The governance of AI research should involve all stakeholders in society. Many attempts have been made to organize the AI safety research community and foster coordinated efforts to develop a more constraining regulatory arrangement for the sector. Independent of the specific policy or measure, the stakes of AI regulation are high, as the global community is collectively at risk of experiencing catastrophic outcomes in which technological advances could go astray. Capturing the multi-dimensional nature of risks associated with AI will be essential to designing AI regulations that are effective, legitimate, and operational [27, 28]. The task at hand is extremely sensitive, as regulations in the absence of cultural acceptance might also engender negative unintended consequences.

8.7.2 Failures and Lessons Learned

The successes of international cooperation in pursuing research and supervising the consecutive development of biotechnology were examined. The same, however, is impossible to find in the sphere of regulation of the military use of the results of this development. Efforts to urge states to accept new legally binding limit and control measures in HL such as proposals to extend the Chemical Weapons Convention type of prohibition to the realm of bioweapons, whose implementation would imply commissioning a public international body that could be given the authority to prevent the dissemination of 'dual-use' molecular machines likely to be used for hostile purposes have led, on the one hand, to division of opinion within the disarmament community, and on the other, to reconfirm the reluctance of national governments to 'suppress' future advances in the biotechnology sphere with a view to allegedly preventing a military misuse of the fruits of this progress [28]. This has triggered a debate about the legitimacy and effectiveness of programs for the prevention of misuse of life sciences. This study is aimed, first, at identifying the situations that have led to the failure of the disarmament process of the biotechnology sector, and second, at exploring the transforming character of failed attempts at saving the current problem of expanding the Chemical Weapons Convention type of regulations to the field of genetically engineered bioweapons. It will argue that today's failed attempts have generated a large number of reassessments of the regulation of the biotech sector for military security purposes, attempting to set a new way and to provide more hope for the future of the control of the military use of biology [29].

8.8 Future Directions in AI Regulation

Because of the early stage of AI development, much remains uncertain about the technology's long-term prospects. The future trajectory of AI systems is likely to depend on both the progress of AI research and development and the pathways by which advanced systems are transferred and commercialized. Building on a framework, we can advance four principles for long-term AI governance:

(i) enhancing transparency for advanced and creative AI systems,
(ii) ensuring careful development of AI design strategies,
(iii) affecting AI collaboration and cooperation, and
(iv) improving near-term AI research and monitoring.

These principles aim to help humanity continually reap the benefits of AI progress and minimize the entropy of AI disasters. Realizing the vision of the AI summer needs further evolution in a rapidly changing technological landscape. In the future, the principles outlined above could be elaborated further. The next generation of AI governance efforts would benefit from collaborative data collection efforts and research-driven deliberative processes to identify the gaps and opportunities for voluntary measures, national regulations, and new international agreements. Additionally, many of the principles constructed here balance the long-standing tensions between the creation of scientific freedom and the need for oversight of dual-use research [30]. AI scientists and policymakers can build their foundations on existing reviews and governance procedures used in the parallel domain of human genetics and, in some cases, proliferate the connections between AI risk management roles and those reserved at the international level for genomics.

8.8.1 Emerging Technologies

Emerging technologies are technologies that are advancing rapidly and have a wide range of potential future uses, most of which are not yet known or well understood. Emerging technologies push the boundaries of what is known by science and often involve never-before-seen applications. We are witnessing this push today in the advance of technologies such as genetic engineering, nanotechnology, quantum information, biorobotics, biometrics, and, of course, artificial intelligence. These technologies will significantly impact our everyday lives, both physically, psychologically, and economically. In the case of artificial intelligence, machines are now capable of carrying out tasks that have, until recently, relied on human intelligence [31]. These may be very narrow tasks or broader capabilities. AI has already been used to perform voice recognition, language translation, sensory focus, advanced robotics, and continues to advance quickly, far more quickly than independent monitors are capable of documenting and understanding the issues associated with that acceleration. The rapidity of advance itself has already called into question whether humans can control the growth of AI towards superhuman AI. Indeed, human control has become a central question for AI scientists who note that the human context for such control no longer involves domination of the natural world by a single approach but expands to include other technologies at work, favoring different outcomes. The development of nanoscale and biotechnology capabilities intersects with the work of AI specialization in the goal of machine superintelligence [32].

8.8.2 Global Cooperation Efforts

In this chapter, a number of efforts on a global level to shape the role and potential of AI are discussed. These efforts cannot be done at the state level alone. AI is an area that exists in an integrated world. Technology and personnel span the globe. Market and commercial interests can reshape the development and deployment of AI systems, often faster than sovereign actions. International agreements, laws, and organizations can help guide AI to benefit humanity. At the same time, ensuring that AI contributes to the greater good needs to consider the distinct legal and political systems that currently comprise the international order. The goals, norms, or guidance may take months to develop, and possibly even longer to be fully implemented on a global scale, but the AI community must start collaborating on these efforts. The established principles are important early steps, but their true value will come from involving officials from a wide variety of governmental and international organizations over the next few years [33]. The entities acting solely or primarily on an international scale that interact with AI include various organizations dealing with a range of policy issues and opportunities related to AI on some level. Additionally, professional organizations or multi-stakeholder groups collaborate with these organizations to develop further frameworks such as the AI Principles. The primary and initial focus of these international efforts concerns AI in the military and criminal space. They are working to establish a more robust understanding of how AI and its enabling technologies are reshaping security dynamics, conflict risks, and defense capabilities. The international organizations are also engaging with appropriate stakeholders and interested parties to establish norms and principles that maintain high standards in security, due process, and respect for human rights in an increasingly AI-driven world. The ultimate goal is for Europe to be the norm setter and an influential actor for AI that is ethical, safe, and obtainable [34].

8.9 Challenges in Regulating AI

We have broached some of the main concerns involved in governing AI and in regulating the multinational corporations that create and widely deploy such systems. In the next and final section, we set out a series of challenges on which both AI companies and the broader public might ponder, given that the smooth operation and strict regulation of all AI systems, not just those that happen to serve one's interest, are in effect common goods. And in order for AI companies and other stakeholders to work toward solutions for the governance dilemmas articulated, we set out some rudimentary principles for experimenting with AI regulation and for sharing the benefits accrued in the short, medium, and long term [35]. This is the groundwork that we propose for establishing AI's governance goals, given where the field is today, both in terms of technology and in its distribution—the latter in terms of where around the world AI's political clout seems to lie. To help good AI regulation take off, what practical steps, working with a mixture of politicians, bureaucrats, businesspeople, and leading thinkers and activists, would these citizens propose? Here are some of the potential next steps and challenges they ventured as part of the pilot experiment, which was largely in line with where participants had hoped to see AI head around a year earlier for the challenges that we have articulated [36]. The politically progressive entrepreneurs, researchers, and artists in the room had made numerous suggestions, some of which have now been further fleshed out to save national electricity accounts and to protect the climate and society from AI's worst wastes and data-hungry applications.

8.9.1 Technical Challenges

As a result of AI posing a myriad of risks to international security, international law, national security strategy, economic security, and social security have all had to change. For this grand challenge, old regulations, which work via largely manual inspection regimes or via detailed written guidance, are less effective. Alas, there is no magic wand to guarantee future safety from AI. It is highly likely that within 15–20 years we will see widespread deployment of AI systems that can forecast how a user's actions will influence much of the content accessible through the internet. But, for the present, the latest developments in AI tend to involve decision systems. From an offshore regulator's point of view, much of the challenge is to create an environment that encourages developers of AI systems to be responsible in their development of these systems, by focusing not only on high performance, but also on the context in which these systems are likely to be deployed. Hence, the question is not whether something could work, but whether society should allow it to be developed and if using it aligns with societal rules, laws, and norms [34]. Therefore, the key question being posed is: At present, what is the likelihood of any AI-controlled system firing a weapon in less than five to ten years respectively?

8.9.2 Political and Economic Barriers

Many political and economic barriers will likely slow the adoption of legal and economic means to control harmful AI systems. A property rights doctrine that is unsuited to many AI systems would be altered absent AIs' capability to generate large industry profits. The many obstacles to criminal deterrence and moral and ethical codes of conduct can only be overcome through patriotism, strong and flexible international treaty systems, or international systems of governance, or some combination of all three. A substitute for regulation in many cases, whether imposed legally or economically, would be the use of AI systems on lethal weapons in the military, security, and policing sectors. Industry, university AI researchers, and many other individuals with high public profiles would likely oppose these mandates [37]. Policy debates would be marred by seemingly endless challenges over controlled studies looking at thwarting external interference to AI systems employed under realistic battle conditions and many other technical aspects of the employment of AI technologies, and investments by the governments of democratic nations.

Because of the potential for disruptive mass violence, the international consequences of failing to agree upon some summation of the behavior that would constitute a violation, and the disparate national interests, treaty systems would need to be structured in such a way as to appeal to self-interest. For government-sponsored actors, only small fines and the likelihood of corporate bodies and third parties assuming liability would foster positive behaviors. Because criminal liability and moral or ethical codes of conduct are viewed as highly unlikely mechanisms for controlling AI catastrophes, it is recommended that future controls be predicated upon economic rewards and equally, if not more severe, detection and procurement systems that are associated with notorious hits [38, 39]. Since the international regulatory process would likely be very slow or immobile, the objective for many AI safety and ethics advocates would be one of doing everything possible over the next ten years or so to raise awareness among AI market participants, with attention directed toward concrete steps that industry leaders and innovators can take collectively to develop economic regulations that are most likely to impact AI as a goal.

8.10 Conclusion

In this chapter, we have reviewed the landscape of institutions involved in the regulation of artificial intelligence and machine learning across many documents and endeavored to extract the main concerns and recommendations from each. Despite the difficulties in this endeavor, we have laid out the viewpoints enunciated in the documents: some see that AI will generally improve humankind's lives and suggest measures to ensure that the newly created wealth will be shared prosperously, while others see that there are fundamental risks to human civilization that come from creating intelligence stronger than humans. Many of our contributors think of ethical principles that should guide the usage of AI and appeal to civil society and research communities to develop and apply these principles. We found very interesting overlaps in the public outlooks of our analyzed institutions and only small differences in conceptualization. We found more marked differences only when going further down in the details. This might reflect a broad convergence of many stakeholders on a small set of big ideas: work on fairness, transparency, accountability, and privacy, using impact assessments and auditable AI. Nevertheless, this outcome should not lead us to downplay the virtues of these emerging ideas about regulation that are increasingly endorsed: if different institutions similarly see things, we should have reason to believe that these proposed measures are more likely to be beneficial in practice.

References

1. Rumpala Y. what about regulating artificial intelligences? On the interest of returning to the fictions of cyberpunk to understand an unresolved challenge. Droit et Societe. 2023;113(1).
2. Shafik W. Artificial intelligence models to prevent forest fires. In: AI and IoT for proactive disaster management. IGI Global; 2024. p. 78–106. https://services.igi-global.com/resolvedoi/resolve.aspx?doi=10.4018/979-8-3693-3896-4.ch005.
3. Karpysheva Y. prosecutors prevention of violations of legislation regulating artificial intelligence: problem statement. Russian J Criminol. 2022;16(3).
4. Goldstein H. AI everywhere, all at once: it's time to get serious about regulating artificial intelligence. vol. 60, IEEE Spectrum. 2023.
5. Tamò-Larrieux A, Guitton C, Mayer S, Lutz C. Regulating for trust: can law establish trust in artificial intelligence? Regul Gov. 2024;18(3).
6. Gutierrez CI, Aguirre A, Uuk R, Boine CC, Franklin M. A proposal for a definition of general purpose artificial intelligence systems. Digit Soc. 2023;2(3).
7. Shafik W. Machine learning for advanced wireless communication: applications, challenges, problems, and open research questions. In: Microwave devices and circuits for advanced wireless communication. CRC Press; 2024. p. 252–78.
8. O'Halloran S, Nowaczyk N. An artificial intelligence approach to regulating systemic risk. Front Artif Intell. 2019;2.
9. Egorova MA, Minbaleev AV, Kozhevina OV, Dufolt A. Main directions of legal regulation of the use of artificial intelligence in the context of a pandemic. Vestnik Sankt-Peterburgskogo Universiteta Pravo. 2021;12(2).
10. Digilina OB, Teslenko IB, Nalbandyan AA. The artificial intelligence: prospects for development and problems of humanization. RUDN J Econ. 2023;31(1).

11. Shafik W. Sustainable development. In: Viana Hassan ASSB, editor. Building community resiliency and sustainability with tourism development, 1st ed. IGI Global; 2024. p. 1–30.
12. Goh HH, Vinuesa R. Regulating artificial-intelligence applications to achieve the sustainable development goals. Discov Sustain. 2021;2(1).
13. Darman R. Peran ChatGPT Sebagai Artificial Intelligence Dalam Menyelesaikan Masalah Pertanahan dengan Metode Studi Kasus dan black box testing. Tunas Agraria. 2024;7(1).
14. Truby J, Brown RD, Ibrahim IA, Parellada OC. A sandbox approach to regulating high-risk artificial intelligence applications. Eur J Risk Regul. 2022;13(2).
15. Manap NA, Abdullah A. Regulating artificial intelligence in Malaysia: the two-tier approach. UUM J Legal Stud. 2020;11(2).
16. Shafik W. An overview of computational modeling and simulations in wireless communication systems. In: Computational modeling and simulation of advanced wireless communication systems. 2024;8–40.
17. Erdélyi OJ, Goldsmith J. Regulating artificial intelligence: proposal for a global solution. Gov Inf Q. 2022;39(4).
18. Shafik W. IoMT Future trends and challenges: emerging technologies, policy implications, and research questions. In: Lightweight digital trust architectures in the internet of medical things (IoMT). 2024;348–70.
19. Bartneck C, Yogeeswaran K, Sibley CG. Personality and demographic correlates of support for regulating artificial intelligence. AI Ethics. 2024;4(2).
20. Higgins DC. OnRAMP for regulating artificial intelligence in medical products. Adv Intell Syst. 2021;3(11).
21. Iphofen R, Kritikos M. Regulating artificial intelligence and robotics: ethics by design in a digital society. Contemp Soc Sci. 2021;16(2).
22. Shafik W. Deep learning impacts in the field of artificial intelligence. In: Deep learning concepts in operations research. New York: Auerbach Publications; 2024. p. 9–26. https://www.taylorfrancis.com/books/9781003433309/chapters/10.1201/9781003433309-2.
23. McLaughlin M. Regulating artificial intelligence in international investment law. J World Invest Trade. 2023;24(2).
24. Shafik W. Generative artificial intelligence for social good and sustainable development. In: Generative AI: disruptive technologies for innovative applications. 2025;153–185.
25. Shafik W. Generative AI for social good and sustainable development. In: Generative AI: current trends and applications. Springer; 2024. p. 185–217.
26. Andryeyev V. Dynamics of regulating artificial intelligence. J Russian Law. 2020;8(3).
27. Martini M. Regulating artificial intelligence—How to de-mystify the alchemy of code? SSRN Electron J. 2019.
28. Ghazmi SF. The urgency of regulating artificial intelligence in online business sector in Indonesia. Hukum Lex Generalis. 2021;2(8).
29. Van der Linden T. Regulating artificial intelligence: please apply existing regulation. Amsterdam Law Forum. 2021;13(3).
30. Gans JS. Self-regulating artificial general intelligence. SSRN Electron J. 2018.
31. Pedram M. Reliability of regulating artificial intelligence to restrain cartelization: a libertarian approach. Asian J Law Econ. 2021;12(2).
32. Harasimiuk DE, Braun T. Regulating artificial intelligence: binary ethics and the law. 2021.
33. Abe O, Eurallyah AJ. Regulating artificial intelligence through a human rights-based approach in Africa. Afr J Legal Stud. 2021;14(4).
34. Mannes A. Governance, risk, and artificial intelligence. AI Mag. 2020;41(1).
35. Shafik W. Community and artificial intelligence-enabled disaster management and preparedness. In: Navigating natural hazards in mountainous topographies: exploring the challenges and opportunities of living. Springer; 2024. p. 243–66. https://link.springer.com/10.1007/978-3-031-65862-4_13.
36. Narechania TN, Sitaraman G. An antimonopoly approach to governing artificial intelligence. SSRN Electron J. 2023.

37. Daćków M. Regulating the unregulatable: EU law an d Artificial Intelligence. Studenckie Prace Prawnicze, Administratywistyczne i Ekonomiczne. 2021;34.
38. Papyshev G, Yarime M. The limitation of ethics-based approaches to regulating artificial intelligence: regulatory gifting in the context of Russia. AI Soc. 2024;39(3).
39. Ellul J, Pace G, McCarthy S, Sammut T, Brockdorff J, Scerri M. Regulating artificial intelligence: a technology regulator's perspective. In: Proceedings of the 18th international conference on artificial intelligence and law, ICAIL 2021. 2021.

Chapter 9
Mitigating the Dark Side: Responsible AI Development and Ethical Solutions

9.1 Introduction

It is important to understand the potential for AI to menace society before discussing how it can be a constructive force. AI could present a threat to society, not just at a micro level, as in the automation of jobs, but also at a macro level. AI systems, if not purposefully directed by humans acting in immediate self-interest, might take steps that could undermine values that are basic to society. AI, as an autonomous system, could propagate and aid pernicious behavior at scales, at speeds, and with capabilities that have not been previously available in human or machine action. This threat is the dark side of AI—an inversion from technologies and systems that have been willingly developed by human society [1]. Responsible AI development must begin with this recognition so that the outcomes driven by AI result in the results intended and valued by human designers—good for the individual, and good for society.

The effects could be felt initially by the displacement of workers and potential harm to individual workers who have the arduous task of re-skilling or changing jobs. But they might be directed with knowledge to benefit the individual while undermining larger system benefits. Even more chilling would be efforts to intentionally work through AI capabilities, designed to take advantage of natural system vulnerabilities, to disrupt or manipulate social, financial, or other systems for competitive advantage or to foster social, political, or other forms of instability. If we wish to preserve autonomy and privacy, know why decisions are being made, and have explanations we can understand, then today's AI is not the solution [2, 3]. The dark side of AI is in contrast to the natural application to aid scientific progress and to support basic tenets enshrined as ethical expectations of the development and use of both human scientific knowledge and technological power.

W. Shafik, *The Dark Side of AI*, https://doi.org/10.1007/978-3-032-09130-7_9

9.1.1 Bias and Discrimination

Automation in decision-making processes powered by AI technology has raised concerns about fairness. In particular, AI is being used to make high-stakes decisions, which at a minimum include hiring, lending, and insurance underwriting. Although AI is much more accurate than human decision-makers in many domains, it is increasingly recognized that AI will replicate both social bias and discrimination from historical hiring, lending, and underwriting data if its development is not tempered with other values, such as fairness and due process. There are several ways in which AI can be unfair in decision-making. For instance, AI creates errors because diverse voices are not included in the development process [4]. Commonly, developers do not consider fairness in their objectives; they focus on performance metrics such as precision and recall. This allows other values, such as equity, to be ignored or unknown. The worst-case scenario arises when the system is catastrophic. For instance, because the training data for credit scoring systems historically have included race as a variable that often takes into account certain practices, having demographic information in the dataset results in credit scoring systems that can be proven to be biased. This exacerbates financial disparities for consumers, who are more likely to face economic difficulties in the future due to these biased predictions. Providing services to low-income groups is also important, but AI fails when decisions are made based on income status [5]. Indeed, as more decision-making services turn to AI for assistance, the equal protection imperative for algorithmic recourse will lead to increased efforts to classify fairness-violating behaviors as fraudulent because those consequences lead to adverse action.

9.1.2 Privacy Concerns

As AI systems raise unparalleled privacy concerns due to the significant amount of personal data involved, these systems create the perfect storm for unintended consequences in the current state of AI development. A few drivers of privacy concerns are the pervasiveness of AI systems in today's world, which collect and process unprecedented large amounts of personal data, the inherent secrecy about the algorithms and data utilized in AI, and finally, the overarching influence these systems could have on society. It is high time for the AI research community, governments, and developers to mitigate privacy concerns in AI systems through responsible development. To promote safe and usable AI, perspectives and expertise in privacy must meld in this time of fast-changing, unexplored possibilities at the crossroads between AI and personal data protection [6]. The shift from traditional programming methods to AI development enhances the influence of privacy. In traditional systems, humans directly code if-else statements or logical rule sets to transform raw input into data in a desired format. However, in AI algorithms such as deep learning, programmers merely feed data to the algorithm, and the system automatically discovers the patterns

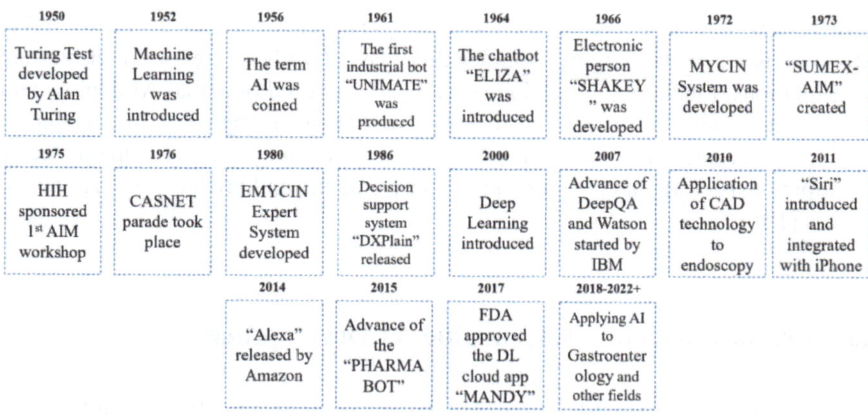

Fig. 9.1 History of AI

by hyper-optimizing. The decision-making process in AI systems is not explicitly programmed: AI systems can detect features and make potentially abstract decisions without human feedback at the time of prediction. Stimulated through exposure to more and more input data, AI constantly optimizes, which can discard most statistical flukes and retain high generalization efficiency overall [7]. Grasping the nature of unexpected outcomes in the AI decision-making process demands deeper probing and opens up wide possibilities for flexibility, since the early 1950s, as presented in Fig. 9.1.

9.1.3 Autonomous Weapons

The fatal nature of weapons, particularly when wielded in warfare, raises unique ethical considerations when they incorporate AI technology. While some focus on the potential for autonomous weapons to lessen human loss by enabling the rise of the 'perfect soldier', critics express opposition on two separate bases. Critics estimate that autonomous weapons will decrease ethical safeguards through the ease and scale of their operational use and raise strong concerns about the ability to ensure the AI programming will not fail or be overridden. Moreover, autonomous weapons, in enabling the digitization of war, may transform the character of warfare and enable a greater inclination towards armed conflict [8]. The U.S., U.K., and other nations oppose any ban on fully autonomous weapons, arguing that existing laws of war can fill the ethical void society fears. However, as the speed of technology awaits no government discussions or negotiations, many other nations and non-governmental organizations see the necessity of unifying efforts to establish international bans or limits on the use of AI technology that can kill without direct human input. As of January 2020, 30 countries and many non-governmental organizations endorsed negotiating legally binding limits on lethal autonomous weapons systems; 26 of these parties sought a prohibition on developing, producing, or acquiring lethal autonomous weapons systems. The deadline passed without creating non-binding AI regulations, much less regulations specifically focused on lethal AI deployment [9]. The increased proliferation of swarms, systems, or recognize-strike technologies, and other autonomous AI weapons makes controlling or eliminating a clearly developing existential threat an urgent matter for public and private organizations to address [10].

9.2 Frameworks for Responsible AI Development

While core principles for the responsible development of AI will help guide developers and stakeholders, elaborated implementations and a supporting infrastructure are needed to ensure that these principles have the intended effect of facilitating AI systems for the common good. We discuss practical guidelines that employ enforceable mechanisms, as well as summarize supporting actions for ethical principle implementation, noting that issues arise that would require legal or industry regulatory oversight. We discuss the ability of these approaches to accommodate the opacity of AI and the ongoing technological development and innovative use of AI across a broad array of domains [11]. We do not interpret that proposed AI regulation is unnecessary, but we aim to provide a higher-level guide for those who wish to steer the development and use of AI, whether or not regulations might apply in their domains of interest. In this section, we provide a horizontal view of the main requirements, regardless of which ethical principles have been chosen to govern AI. In order to propose positive steps for ethical development, we start with a proposal of high-level requirements for responsibility in AI development. Where possible, we follow basic engineering principles, complementing these with insights from risk management and the economics of market and information failures, and by promoting positive economic and societal incentives [12]. We also attempt to balance what can be contradictory objectives, such as the lack of limits in creative endeavors, the need for oversight, and the desire for human review and explanation functionalities in certain AI developments.

9.2.1 Ethical Guidelines

A further form of beneficial constraint involves the development and application of ethical guidelines for AI and autonomous systems development and use. These guidelines could be fertile ground for academics in the fields of philosophy, computer science, law, and others. Guiding principles might include adherence to principles of beneficence, care, nonmaleficence, and respect for autonomy in AI and autonomous systems design, development, and applications [13]. These principles are akin to the moral views of the collected authors. The principle of beneficence entails a human's obligation to assist others in need. The principle of care explains that people may take special obligations to care for those with whom they have close and personal relationships [14]. The principle of nonmaleficence is the familiar 'do not harm.' The principle of respect for autonomy entails treating individuals as cognitively and emotionally competent members of the moral community.

It could be that both soft and hard constraints would need to be advocated to prevent the development of autonomous intelligent robots whose use would potentially yield a 'bad life' for their users in the Aristotelian sense. Since the technology and its deployment are in their early stages, individuals from philosophical, legal, and other fields might create soft law regulations that pick up on the virtue ethical perspective. These regulations might initially focus on constrained but high-level functions, such as remote surveillance or lethal force. However, it is doubtful whether such hard and soft regulations would suffice [15]. The principle of nonmaleficence may need hard regulatory support to ensure the 'not harm' maxim. The challenge is that autonomous intelligent robots capable of forming a 'robust moral compass' are some years in the future at best. As a consequence, we may have to shield ourselves from ourselves. The logic of this is to surround moral AI, the initial ability for a machine to make ethical decisions, with strict and strong laws and regulations that prevent the physical or moral harming of people. We would be prudent to minimize the presence of risk faced by both the initial developers of the autonomous artificial general intelligence and the robotic system's potential users [16].

9.2.2 Regulatory Frameworks

Earlier, we mentioned that in an evolving, broad, and nebulous field, heavy-handed or too specific regulation could limit the field's potential, impair development for ethical uses and penalties, and prevent one-off uses to protect the environment or other needed steps. Although we agree, we also think that this does not mean that the field of artificial intelligence would not need some regulation. To guide developers of strong AI systems, at the least to help keep in mind the risks and ethical issues, we think that the recommendations mentioned have sustaining value [17]. For the moment, we assume that the existing self-regulatory or research and development community guidance contingent would surely rather that their guidance was based on some framework. Ultimately, however, we may expect the lightweight, collaborative research among facilities responsive to an interinstitutional conventions approach to de-emphasize the tendency among collaborations, as well, and lean more toward collaboration with regulation for safety. In this spirit, we suggest that this would be an opportune time to attach 'privacy', 'fairness and welfare', and 'human control of artificial intelligence' provisos to any safety-oriented recommendations that we find to pass a reasonable 'strong AI vulnerability assessment' [18].

9.3 Stakeholders in AI Ethics

The development of AI should consider the present as an active exchange between all stakeholders, including IMS. The ultimate goal of this exchange on AI and ethics is to encourage convergence toward internationally agreed-upon standards underwritten by shared values. Global support is needed to help resolve divergent interests regarding specific values and comprehensive global standards. Representatives from many important fields must actively participate in an ongoing and mutually respectful discussion about challenges associated with the development and use of AI. However, at its very core, the community of ethicists must play the role of defending the values of human rights and democratic principles when these values are threatened [19]. In recognition of the rapid and far-reaching technical change that IMS is currently witnessing, public policies and new technologies do not sufficiently consider the scale of the ethical problems and the limits of available solutions. In light of the urgent need for a strong community of knowledgeable practitioners, the community consists of technical experts in policy, law, and ethics who can provide practical value to IMS customers. Educational scientists in this area must recognize that the ethical component is an essential part of graduation. As key stakeholders in engineering education, employers and recruiters have an important role to play in the development of educational standards by helping to align educational offerings with the needs of the industry they serve [1]. They also need to articulate the development of ethical criteria for engineering practice with respect to future technologies that may not yet exist.

9.3.1 *Governments*

The bigger issues revolve around greater societal and institutional responsibilities associated with ensuring that AI benefits all. Governments have a critical role in setting the rules of the game—that is, establishing the legal and regulatory frameworks to ensure societal and environmental liabilities are internalized by the institution producing AI. Establishing the right balance of public and private sector production of AI to ensure the interests of all sectors are reflected is similarly important, enabling the full benefits of AI, beneficial for all parts of society [4]. Governments are also critical in establishing the legal frameworks that ensure the rights and obligations of all stakeholders are balanced, along with their enforcement and oversight. It is critical to establish international norms and standards that reflect societal and environmental risks and contain them as well. International collaboration reflecting multi-stakeholder agreements informs responsible development and use of technology, the roles and obligations of multi-technology companies, and agreed research and development investment to avoid costly and unwanted technology conflict and promote technology diffusion [3].

9.3.2 *Industry Leaders*

Heavy hitters in AI development have begun efforts to tie artificial intelligence to the rules of ethics. Recognition has finally dawned within leading companies, and an impressive approach addressing this issue is growing and increasingly formidable. One company takes the initiative of serious mitigation moves in the following general approach [20]. First off, while acknowledging the dark side of this humanlike intelligence-craver, this company deems industrial AI as an extension and expansion of human abilities, augmented intelligence with all the mandatory ethical constraints to which its human model is subject, certainly not artificial intelligence, but human intelligence in asymmetric form. In addition, they provide not as banning prohibitions, rigid corporate checkpoints, or advanced correctness tests, but as advisory, guiding hotlines on moral issues [2]. These continuously updated guides help to decode and interpret the rules of business to balance the enthusiastic welcome embrace of tremendous profits with the weight of due respect for the appropriate positioning of alterations to the form of human intelligence deserving of concern and respect.

9.3.3 *Academics and Researchers*

Academics and researchers help promote a socially responsible and ethical use of AI by continuously improving public knowledge about the opportunities and risks of AI technologies, and by training and setting best practices for the management of these technologies. That is the mission of numerous research centers and the objective of numerous research projects and scientific publications. Just as the researchers' work related to the development of AI is not new, and mainly expands in institutionalized form with the creation of the first research laboratories at the beginning of the 1960s, this impulse to the mission of ethicizing AI also has its recent antecedents [21]. Since 2017, two independent research institutes have published annual agenda-building reports with considerations and multidisciplinary recommendations, both for researchers in the development of AI, as well as for decision-makers and the general population. The scientific production seeks to be open, in open access, and does not seek profit or intellectual property protection. As a place of critical thought and construction of ethical norms, it deserves to be fostered and heard, as sampled tools are presented in Fig. 9.2.

The importance of these institutes' mission and, in general, of the researchers published in journals on scientific dissemination lies in the fact that research development is intrinsic to the operation of AI technologies. This action includes, but is not limited to, the choice of the theoretical basis and development of algorithms and architectures for decision-making of AI lifecycle management teams and investments [22]. The understanding by organizations and the general public of the impacts, control, and interaction with these that AI can generate requires tolerance with the

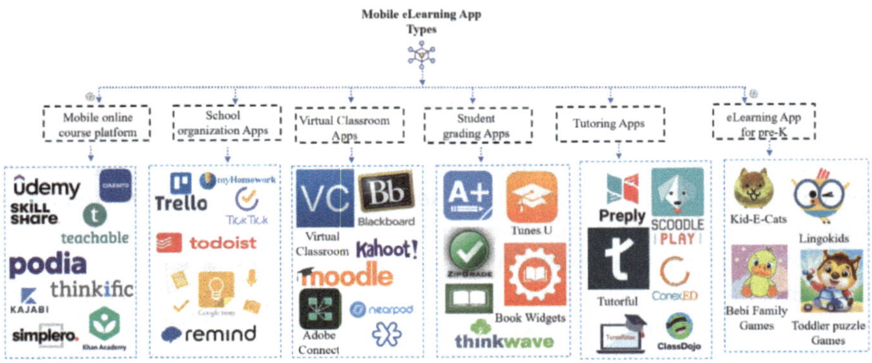

Fig. 9.2 Mobile eLearning applications

limitations and capacity for intervention and feedback of AI. The technological analysis efforts required, in open and ethical ways, to mitigate, promote, and regulate the benefits and risks of the implementation of these technologies are advanced and directed to the forecast of these impacts. In summary, the self-perception and the development of capabilities require research with transparency, self-criticism, future vision, and long-term values [23].

9.3.4 Civil Society

A unique layer in the ecosystem of AI is the involvement of several organizations from civil society. The wave of AI "ethics washing" shifting to a constructive critique is fueled by the work these organizations are doing, collaborating with researchers, the technology industry, and the public sector. Civil society groups are working on technical and policy issues, highlighting the social and ethical aspects of AI, and at the same time, they are in the field as partners with technology developers implementing AI for social good and addressing societal issues [24]. Initiatives such as various technology companies, civil liberties organizations, and many others, together with think tanks, academic institutions, non-profits, and activist organizations, all share a common interest within the civil society ecosystem, striving to bring forth a collective engagement with AI development. With regard to advocacy around decolonization and diversity, and as the use and deployment of AI become the domain of the criminal justice system, media, and other lenses that raise controversies, indeed, civil society organizations point to multiple unavoidable and pertinent questions. AI technology products and solutions built through civil society-led research and recommendations are seen as the most trusted and responsible ones [25].

9.4 Case Studies of Ethical AI Implementation

To make the concrete application of these case studies more transparent, we can analyze a few case studies, reflecting not only some of the most challenging processes and tasks but also some of the most ethically and technically difficult principles. The first case study is that of an AI for carbon footprint measurements. Recognizing the importance of these goals with some of the most cutting-edge models, the aim is for the reliable prediction of the allocation of energy used in particular hardware devices in a task, given the diversity of their possible characteristics and the states of a given hardware device during the processing of a given task [26]. In the second case study, the AI that is fulfilling the function of pricing products, which is capable of carrying out a lifetime of work in a split second, will make decisions with the most profound ethical implications for people across its market geographies. Questions arise about how the intrinsic properties of AI, derived from the principles and ethical framework from which it has been built, can still be integrated into a framework of social values and regulatory obligations. The two studies come from well-known applications and objectives in the AI and corporate world [18]. The first case study is a dedicated study: AI for specific purposes—a machine learning model built to predict the carbon footprint of device-specific code usage in factories.

9.4.1 Successful AI Projects

A successful AI project requires informed, multidisciplinary professionals, with representation from business units tasked with mapping out important ethical considerations for that function. Although AI models are data-driven and do not theoretically require human input, stopping unethical behavior remains necessary. The possibility of reaching conclusions that are reasonably free of bias requires testing models to determine which variables are the most important. Results should be regularly reviewed and interpreted by human domain experts in order to provide insight into decision-making behavior and facilitate further testing of machine learning techniques over time [16]. In the absence of attention from domain experts, the typical opportunity cost will be the failure to derive substantive conclusions from embodied biases.

Whether or not AI validates a researcher's hypothesis, human feedback remains crucial. It is important, using data, to weigh the relative performance of an AI approach. Practitioners may need to aggregate less accurate AI models with other models that use different methods when a technology solution is difficult for them to construct, while requiring a high level of understanding and judgment. Introducing deep learning in a low-frequency finance context might lead to a forecast architecture that hardly predicts at all but overweighs the tiniest subpopulation with the inability to assess risk [27]. Unfortunately, in the process, it may not achieve the typical result of the lowest aggregated error. Such an approach would lack naturally efficient elements and not adequately represent the relevant segments of the distribution of predicted outcomes, the way that human judgment might. These procedures require still further data analysis. Ideal models today might not exist for use in a variety of domains [14, 15]. Custom, hand-designed AI methods may consistently outperform configurable package AI methods for the potential domains of use.

9.4.2 Failures and Lessons Learned

At their core, most AI mishaps are human mishaps, arising due to unmet obligations and inattention to details. Failures are exacerbated by organizational pressures regarding cost, timeliness, and performance, tempting developers to take shortcuts. Such missteps run the risk of technical error, leading to inadvertent harm, and ethical error, breaching community norms. By taking a deep look at those examples, some general lessons emerge. First, inattention has led to development errors that don't fully consider the potential for undesirable bias. Learning systems driven by an economy of data processing from sources prone to bad training data and data-mined decision support can reflect all the human biases and errors that have gone before us [12]. Yet the makers of these systems can remain inattentive, falling victim to the semi-black-box nature of these solutions.

Quite a list of questions is hard for outsiders to answer, but it's incumbent on the developers of AI success story use cases to put them into a standard checklist. This is particularly true because such technologies can then evolve into industrial-scale deployments, with a small development team responsible for operating widespread systems [28]. The most intense pressure today involves securing machine learning against a realistic adversary, or focusing the performance of such systems on a narrow task where model performance is reliably high and failure modes can be anticipated and handled straightforwardly. We must guard against the current unmet obligations arising from successful research transitions manifesting as ethical error [9]. To execute political will in development tasks and yield deployment sufficiently sensitive to ethical concerns, we must get good at translating principles.

9.5 Technological Solutions to Mitigate Risks

As AI use evolves out of early stages, robust solutions are essential to protect privacy rights, prevent biased decision-making, and ensure safety in applications that involve important decisions, such as in criminal justice and healthcare, as well as in broader commercial and national security contexts. Concentration on agility generally drives technology developers to focus on creating novel machine learning algorithms and general-purpose services. This goal is appropriate; rapid research to develop faster and more efficient algorithmic training and inferencing reduces costs and prevents drawing inappropriate conclusions during development [8]. But avoiding inappropriate focus on only the technology's outcome, rather than on the effects in different contexts, creates critical gaps later in the development process, in the context implementation design phase. Rapid research in this area is also essential; identifying and mitigating the dark side consequences, especially in public sector uses, prevents project delays that can sabotage rollouts. Yet stakeholders in initial uses involving basic and applied research, or public or private commercial development, often evade deep discussion of these consequences. Publication of preliminary results or successful implementations without complete details speeds up new research and project speed, and funds additional related work [7].

To address these issues, developers should plan for appropriate responsibility and ethical solutions, as well as for performance monitoring. Project development should maximize ethical solutions and privacy, and safety protections through a robust technology stage. However, some AI code may present threats regardless of the best-designed models. Therefore, during the development of potential AI applications for national security contexts, AI enterprises should simultaneously create research programs to defend application outputs from algorithmic overconfidence concerns [6]. Addressing inherently unreliable or adversarial machine learning systems at this stage may uncover useful defensive elements that can be embedded early on, before the models are pushed into new contexts. Focused research supports the development of inherent weaknesses that attackers can exploit explicitly, briefly, and metonymically. Individuals may be able to manipulate adversarial models by a change barely noticeable to human perception [29]. Control over the circumstances gives the wrongdoer an advantage. Defenders can prepare in advance, proofing against adversarial, exploitable AI using game theory and anticipated approaches to weakening. In developing models to be secure, this research extends from the preliminary stage, enabling the advancement of almost two o'clock-phase solutions of technology readiness [6]. Adjusting problem framing and prediction goals is the most robust preliminary mitigation strategy to prevent attacks.

9.5.1 Transparency Tools

Open Information Models. Machine learning models can reveal how they work. The Open Information Model lets users understand what changes in input and output layers and model parameters mean. Educators utilize it in their courses, employers trust the models they employ, and regulators permit them for critical tasks, while researchers can bring the models to perform even better. State-of-the-art machine learning algorithms are increasingly based on large models trained on large datasets. To fully realize their promise, they need to be trained on the full range of data, from hundreds of millions or billions of people [9]. As models become too big to easily, privately, and justly store on-device training data, novel combinations of privacy-enhancing technologies are needed to continue the rise of machine learning while ensuring that those who have provided their data have control over what is learned [30].

9.5.2 Accountability Mechanisms

Accountability Mechanisms. Achieving accountability in AI systems is challenging. Many recent works on AI and ethics include a section on the importance of making AI and its outcomes accountable and provide various examples of how this could be achieved in practice. Most of these concern AI design and deployment. However, AI solutions also need to address identified failures and recognize new learning responsibilities. This section provides reasoning for being accountable in AI design and for user-compensating AI failures to mitigate key ethical issues resulting from using AI [6]. AI systems can be held accountable for failure only if users obtain support in handling new responsibilities and if the current accountability deficit of harmful AI is addressed. Post-deployment Responsibility for AI Failure. AI systems should be considered by their developers to be in use for the lifetime of the systems, with time-shared post-deployment responsibility and post-mortem learning. Accountability is defined as the understanding that each of us is answerable to a higher power and that the higher power is the polity as a whole or those damaged by misconduct [31]. Thus, AI decisions need to be transparent, and the outcomes need to be accountable. Instead, AI users are forced to accept AI when the government, business, and services require it, but then the designers escape their legal duty when AI fails. AI service terms typically state that using the AI is the user's responsibility, and its operation may fail. If the user does not agree, then they cannot use the service [5]. On these one-sided terms, the risk of AI failure is moved to the user who has the authority to check or question the service or to compensate for discovered problems, as presented in Table 9.1.

9.6 Public Perception and Trust in AI

Recent surveys have indicated that while a majority of Americans are excited about the prospects that AI technology may offer, a larger percentage of the public is concerned. Uneasiness about the impact of AI on jobs ranks relatively lower compared to other highlighted concerns from recent surveys, such as issues of data privacy, the potential for AI systems to make unfair and biased decisions, AI-led surveillance, or potential misuses of AI technology. Despite this support for technology development and its application, public trust in AI is low, and many believe that AI is being deployed in society without adequate safeguards or oversight. This may be the consequence of sensational news stories that expose various design, variance, and excessive usage limitations of current AI systems [17]. Such stories are woven into a larger public narrative around the power of AI technology, aligning with

Table 9.1 Ethical strategies for mitigating the risks of artificial intelligence

Ethical principle	AI concern addressed	Implementation strategy	Responsible stakeholder	Example in practice	Expected outcome
Transparency	Black-box decision-making	Explainable AI (XAI) models	Developers, researchers	IBM's AI Explainability 360 toolkit	Improved user trust and system interpretability
Fairness	Algorithmic bias and inequality	Bias detection and data auditing	Data scientists, policy makers	Google's PAIR (People + AI Research)	Equitable outcomes across diverse groups
Accountability	Unethical AI use or failure	Clear responsibility assignment	Organizations, regulators	AI incident reporting frameworks	Stronger compliance and safety culture
Privacy protection	Data misuse and surveillance	Differential privacy and encryption	Tech companies, lawmakers	Apple's on-device learning	Enhanced data security and user confidence
Human oversight	Automation without checks	Human-in-the-loop systems	System designers	Autonomous driving requires human intervention	Reduced error in critical decision-making
Inclusivity	Marginalization in AI outcomes	Diverse teams and stakeholder input	HR, developers	Microsoft's inclusive design principles	Broader societal benefits from AI solutions
Environmental sustainability	AI's carbon footprint	Energy-efficient algorithms	AI researchers, infrastructure teams	Google's use of AI for data center optimization	Reduced environmental impact
Open collaboration	Fragmented AI ethics standards	Cross-disciplinary research and frameworks	Academics, governments	The Partnership on AI	Unified ethical standards and shared learning
Regulation & policy	Unchecked AI deployment	Government policies and international norms	Policymakers, international bodies	EU AI Act	Safer and harmonized global AI practices
Public awareness	Misinformation and misuse	Digital literacy and transparency campaigns	NGOs, educational institutions	AI ethics courses in universities	Informed citizens and ethical AI consumption

personal experiences, philosophical and ideological understanding, and the historical context of public threats and risks. The information reported in the surveys can be leveraged to define the boundaries for more responsible behavior when developing, integrating, and utilizing AI. A logical consequence is that AI systems ought to be held to higher standards when addressing risks associated with privacy, fairness, and accountability while achieving robust and reliable performance to ensure veracity and reported accuracy across the wide spectrum of endeavors in which these systems operate. This also calls for a sustained prioritization of regulatory, oversight, and standardization efforts, which are, to date, underdeveloped in the United States and other major AI-powered economies [32]. Moreover, the results indicate that a rapid response and overcorrection by AI developers through the use of techniques could offset the baseline public perception and perception of large technology companies.

9.6.1 Building Trust

The validity of these findings depends, among other things, on the quality of AI models and their ethical underpinnings. AI's potential for creating public value is vast, as are the risks. For AI to deliver positive social outcomes, decision makers must have confidence in the fairness, safety, reliability, explainability, and privacy of these models. A competitiveness agenda that promotes effective management of data, strengthens investments in AI research, and secures talent and skills is critical to create an environment that fosters responsible models of AI innovation and development [20, 33]. AI development processes must engage with objective stakeholders from the public, private, and civil society sectors to best ensure the benefits of AI innovation transcend societies and economies, are widely shared, and serve the public interest. Technology transition must focus on delivering real public value in areas such as health, agriculture, and climate change. The AI age's productivity premium and national security and defense objectives depend on the responsible use of technology. To build trust in the use of AI by government and across society, leading by example and developing AI capabilities responsibly are essential [34].

9.6.2 Public Engagement Strategies

Throughout this article, we have emphasized the importance of public engagement with AI systems. But how do we avoid this simply becoming a tick-box for dark AI development projects? At a minimum, this should involve an understanding of the many groups (and individuals) that are potentially affected by AI solutions, as well as any power and knowledge asymmetries between them. This means considering how values are implemented within AI systems, what potential limitations exist, and what trade-offs are present, both at the code level and at the level of use. Values are not neutral and are relative to the situation in which they are applied. This requires dealing with public controversies and differing views [35]. In considering questions of responsibility, we have noted that trade-offs need to be made between multiple non-moral values. These might include privacy, safety, non-discrimination, and fairness, to name but a few. That is to say, many stakeholders have legitimate claims that are not expressed by moral values alone; these must be justified in terms of non-moral values or resolved through negotiated ethical positions [36]. It is precisely because these differences in views are unlikely to be resolved by appealing to general principles that public engagement is encouraged. Furthermore, in addition to creating and implementing public engagement strategies, we believe we also need to develop care, humility, and empathy for the world in which our dark AI from light AI desires may have unanticipated effects. This means a willingness to make our AI solutions transparent and subject to scrutiny. The crafting of our ethical position and its implementation need to be incremental, in part determined by the level of problem consequence as well as partly due to alternative non-moral values and different attitudes [37].

9.7 Future Directions in AI Ethics

Responsible AI requires not just the management of risk today, but the ability to adapt to changing social and technological landscapes over time. In this chapter, we present a vision for ethical AI that emphasizes ethical solutions to specific problems, trustworthy oversight, and resilience in the face of change. The first two parts of our collective AI ethics approach draw heavily from our work with commercial partners; the last part is inherently speculative. We hope that, regardless of errors in our suggestion, this approach or something like it will beget a generation of machines that we can rely on to serve our values. Recent successes and failures in AI have been met with both awe and alarm [38]. Almost weekly, popular media features exciting developments in AI that promise everything from better medical diagnoses and precision agriculture to a fully automated future of work and a transformed wealth of nations. Other reports caution that these innovations make decisions that are unfair and biased, unsafe and unpredictable, and not respectful of important human values like privacy, freedom, and dignity. To make AI safer, fairer, and more accountable, stakeholders have weighed in with principles, guidelines, and best practices along with calls for stricter regulation, more thorough testing, and better documentation. These efforts have focused primarily on liability and risk management: fairness through process constraints, opacity through transparent accountability, and privacy through secure isolation. However, responsible AI requires not just the management of risk but the ability to adapt to changing social and technological landscapes over time [39]. In this chapter, we present a vision for AI that emphasizes ethical solutions to specific problems, trustworthy oversight, and resilience in the face of change.

9.7.1 Emerging Technologies

Using newer capabilities applied to new domains means that the models and approaches have not been pressure-tested and have led to the emergence of new nefarious behaviors and incentives. As time and technology progress, incentives evolve, making actions more than algorithmically detrimental. This is the dynamic not tackled by holding algorithms and the model creators to ethical standards. We live in a society that operates not by algorithm, yet holds individuals protesting the algorithmic prioritization of COVID vaccine delivery as harmful agents of society. And so we allege that the identity of the individual does not determine the harms, and the harms are the same as the harms being expressed by those ignored by the algorithmic allocation [40]. Artificial Intelligence and Machine Learning are rapidly evolving fields that draw on novel methods for processing information and making decisions. Many new applications are still crude in their operations and affect human life and well-being. This is the case with many applications across different domains of AI/ML research, including medical prediction, hiring, firing, risk management, and autonomous systems in defense, to name a few. Ensuring that these algorithms comply with ethical principles and fundamental human rights guidelines is a high priority to avoid harmful effects that we will then struggle to mitigate. Moreover, as the underlying technology evolves, so too may the predictions of the model, altering the returns and incentives of competing decision strategies in the black-box competition [1]. Such indirect effects, and not just the potential harm from the decision itself, are particularly harmful because those incentivized to produce model understanding truly are not the model beneficiaries.

9.7.2 Global Collaboration

The global nature of the AI field and the potentially non-containable risks of AI call for a sustained global conversation around AI and its governance. Global collaborative efforts can help to disambiguate potential flash points and areas of misunderstanding among diverse national perspectives on what is needed to steer towards beneficial advanced AI. These efforts could be collaborative technical research, public–private initiative building, or platform development efforts to build foundational global governance infrastructure. As AI policymakers are acting with urgency, taking internal national viewpoints as imperatives, building on diverse governance design can also be helpful [41, 42]. Ethical frameworks around AI hold promise, and we believe that robust, open, multi-stakeholder global initiatives could yield governance designs that genuinely harness AI to improve outcomes. An open initiative of this type could disambiguate potential controversies or a lack of alignment among diverse national perspectives on what is needed to achieve that. Such an open process might also yield powerful integrative global AI best practices that improve the alignment of national practices not only with one another but also with some objective standards [43]. With this input, nation-based governance and oversight organizations could help stakeholders to ensure that beneficial AI designs are most likely to be pursued in each country and sector.

9.8 Conclusion

The benefits that AI can produce are manifold. Tasks that used to be done by humans can now be done by AI systems much more quickly and efficiently. However, these new AI technologies also pose new risks and amplify existing challenges. The development of AI technologies, therefore, also involves ethical concerns. When AI-based systems are deployed, various parties are involved, including the system's developer, operator, and end-user. Responsibility involves identifying, understanding, and efficiently bringing to account the involved parties, but also remedying a potential vulnerability and designing for prevention. Fairness and accountability in AI-based systems are more likely to be achieved when designers understand and implement best technical practices, thereby enabling the public to be agents of a democracy in which privacy protection is the means, not the end, of autonomy. AI systems will inevitably make decisions that challenge the status quo. When they do, relying on well-designed AI systems can help avoid creating a false sense of security that allows negative consequences to quietly accumulate behind the semblance of rigorous risk assessment, screening, and decision-making. Sound decision-making is critical for ensuring the quality of AI systems designed to undergird functions as varied as protecting the privacy of vulnerable people and easing the inherent tension between civilian and militarized responses. Developing effective and ethical AI systems that are truly "responsible" begins by identifying leaders who are willing to work through the challenges identified herein. Policymakers must listen, learn, and follow through by supporting safe, ethical, and responsible AI, recognizing that the process may require a fundamental policy shift. The opportunity to protect American consumers and businesses should not be set aside by default or forgo the value that AI developments can responsibly produce.

References

1. Mikalef P, Conboy K, Lundström JE, Popovič A. Thinking responsibly about responsible AI and 'the dark side' of AI. Euro J Inf Syst. 2022;31.
2. Ojha NK, Pandita A, Ramkumar J. Cyber security challenges and dark side of AI. 2024.
3. Belanche D, Belk RW, Casaló LV, Flavián C. The dark side of artificial intelligence in services. Serv Ind J. 2024;44(3–4).
4. Xiao L, Shen XL, Cheng X. Introduction to the HICSS Minitrack "The Dark Sides of AI." In: Vols. 2022-January, Proceedings of the annual Hawaii international conference on system sciences. 2022.
5. Castillo D, Canhoto AI, Said E. The dark side of AI-powered service interactions: exploring the process of co-destruction from the customer perspective. Serv Ind J. 2021;41(13–14).

6. Gligor DM, Pillai KG, Golgeci I. Theorizing the dark side of business-to-business relationships in the era of AI, big data, and blockchain. J Bus Res. 2021;133.
7. Papagiannidis E, Mikalef P, Conboy K, Van de Wetering R. Uncovering the dark side of AI-based decision-making: a case study in a B2B context. Ind Mark Manag. 2023;115.
8. Barman D, Guo Z, Conlan O. The dark side of language models: exploring the potential of LLMs in multimedia disinformation generation and dissemination. Mach Learn Appl. 2024;16.
9. Rana NP, Chatterjee S, Dwivedi YK, Akter S. Understanding dark side of artificial intelligence (AI) integrated business analytics: assessing firm's operational inefficiency and competitiveness. Euro J Inf Syst. 2022;31(3).
10. Shafik W. Cyber security perspectives in public spaces: drone case study. In: Handbook of research on cybersecurity risk in contemporary business systems. 2023.
11. Pantano E, Marikyan D, Papagiannidis S. The dark side of artificial intelligence for industrial marketing management: threats and risks of AI adoption. Ind Mark Manag. 2024;116.
12. Cao L, Chen C, Dong X, Wang M, Qin X. The dark side of AI identity: investigating when and why AI identity entitles unethical behavior. Comput Human Behav. 2023;143.
13. Shafik W, Matinkhah SM, Ghasemzadeh M. Theoretical understanding of deep learning in UAV biomedical engineering technologies analysis. SN Comput Sci. 2020;1(6).
14. Zhou Y, Wang L, Chen W. The dark side of AI-enabled HRM on employees based on AI algorithmic features. J Organ Change Manag. 2023;36(7).
15. Begou N, Vinoy J, Duda A, Korczynski M. Exploring the dark side of AI: advanced phishing attack design and deployment using ChatGPT. In: 2023 IEEE conference on communications and network security, CNS 2023. 2023.
16. Xing Y, Yu L, Zhang JZ, Zheng LJ. Uncovering the dark side of artificial intelligence in electronic markets: a systematic literature review. J Organ End User Comput. 2023;35(1).
17. Wiederhold BK. The dark side of the digital age: how to address cyberbullying among adolescents. Cyberpsychol Behav Soc Netw. 2024;27(3).
18. Alawida M, Abu Shawar B, Abiodun OI, Mehmood A, Omolara AE, Al Hwaitat AK. Unveiling the dark side of ChatGPT: exploring cyberattacks and enhancing user awareness. Information (Switzerland). 2024;15(1).
19. Cheng X, Lin X, Shen XL, Zarifis A, Mou J. The dark sides of AI. Electron Markets. 2022;32.
20. Shafik W. Security, 15 Privacy, and Trust in Fintech. In: FinTech and financial inclusion: leveraging digital finance for economic empowerment and sustainable growth. 2025;216.
21. Pearson A. Refrigeration applications column the dark side of AI. ASHRAE J. 2023;65.
22. The sunny and the dark side of AI. Economist (United Kingdom). 2018;414.
23. Wirtz BW, Weyerer JC, Sturm BJ. The dark sides of artificial intelligence: an integrated AI governance framework for public administration. Int J Public Adm. 2020;43(9).
24. Kumar G, Singh G, Bhatanagar V, Jyoti K. Scary dark side of artificial intelligence: a perilous contrivance to mankind. Human Soc Sci Rev. 2019;7(5).
25. Asif M, Lodhi RN, Sarwar F, Ashfaq M. Dark side whitewashes the benefits of FinTech innovations: a bibliometric overview. Int J Bank Mark. 2024;42(1).
26. Sun Y, Li S, Yu L. The dark sides of AI personal assistant: effects of service failure on user continuance intention. Electron Markets. 2022;32(1).
27. Shafik W. Mapping flood hazards in Sub-Saharan African Region: hindering regional sustainable development. In: Geostatistical insights on mapping flood hazards and wetland dynamics. IGI Global Scientific Publishing; 2025. p. 1–30.
28. Shafik W. Factoring 6G technology and beyond in advancing human life management and natural habitats. In: 6G impacts on natural habitats and human life. IGI Global Scientific Publishing; 2025. p. 319–58.
29. Shafik W. Human-artificial intelligence collaborations in polycystic ovary syndrome (PCOS) clinical trials and research. In: AI-based nutritional intervention in polycystic ovary syndrome (PCOS). Springer; 2025. p. 307–30.
30. Shafik W, Singh R, Kumar V. artificial intelligence transparency and explainability in sustainable healthcare. In: Transforming healthcare sector through artificial intelligence and environmental sustainability. Springer; 2025. p. 165–91.

31. Singh R, Shafik W, Crowther D, Kumar V, editors. Transforming healthcare sector through artificial intelligence and environmental sustainability, vol. 1, 1st ed. Singapore: Springer Nature Singapore; 2024 [cited 2025 Feb 27]. https://link.springer.com/book/https://doi.org/10.1007/978-981-97-9555-0.
32. Shafik W. Ethical and legal considerations in artificial intelligence. In: AI-enabled threat intelligence and cyber risk assessment. 2025. p. 90.
33. Shafik W. Towards trustworthy and explainable AI educational systems. In 2024. p. 17–41.
34. Truby J, Brown RD, Ibrahim IA, Parellada OC. A sandbox approach to regulating high-risk artificial intelligence applications. Euro J Risk Regul. 2022;13(2).
35. Tamò-Larrieux A, Guitton C, Mayer S, Lutz C. Regulating for trust: can law establish trust in artificial intelligence? Regul Gov. 2024;18(3).
36. Pieters W. Explanation and trust: what to tell the user in security and AI? Ethics Inf Technol. 2011;13(1).
37. Lampou E, Antonopoulos N. Ranked by truth metrics: a new communication method approach, on crowd-sourced fact-checking platforms for journalistic and social media content. Stud Media Commun. 2023;11(6).
38. Dash B. Zero-trust architecture (ZTA): designing an AI-powered cloud security framework for LLMs' black box problems. SSRN Electron J. 2024.
39. Diez-Gracia A, Sánchez-García P, Martín-Román J. Disintermediation and disinformation as a political strategy: use of AI to analyze fake news as Trump's rhetorical resource on Twitter. Profesional de la Informacion. 2023;32(5).
40. Alaziz SN, Alshowiman AA, Albayati B, El-Bagoury A al AH, Shafik W. Clustering of COVID-19 multi-time series-based k-means and PCA with forecasting. Int J Data Warehousing Min. 2023;19(3).
41. Shafik W. Generative AI for social good and sustainable development. In: Generative AI: current trends and applications. Springer; 2024. p. 185–217.
42. Shafik W. Deep learning impacts in the field of artificial intelligence. In: Deep learning concepts in operations research. New York: Auerbach Publications; 2024. p. 9–26. https://www.taylorfrancis.com/books/9781003433309/chapters/https://doi.org/10.1201/9781003433309-2.
43. Karpysheva Y. Prosecutors prevention of violations of legislation regulating artificial intelligence: problem statement. Russian J Criminol. 2022;16(3).

Chapter 10
Charting a Path Toward a Responsible and Human-Centered AI Future

10.1 Introduction

Artificial intelligence is the development of computer systems that are able to perform tasks that normally require human intelligence. AI can take many different forms. Machine learning, in particular, has become a dominant approach to AI. Machine learning is a family of algorithms that attempt to learn information directly from the data [1]. In so doing, it can power many of the AI features we see in the world around us today. For example, the technology that has mastered a variety of games, such as checkers and chess, has ultimately learning to beat the best humans. In the process, it actually learned a lot from data, although the data in this case came from playing against itself. Machine learning has been as much of a breakthrough as it has because it allows companies to build systems that can learn from data, rather than being programmed explicitly for certain tasks [2]. This allows systems to learn from the large amounts of data available today and to be carefully tuned to the task at hand. The boom in computing power has enabled machine learning to become a powerful tool. Through the use of many layers of interconnected processing units and algorithms to learn to adjust their connection weights, deep learning can learn deep hierarchical representations of data. This allows the technology to learn the structures behind the data. Such systems have transformed industries by automating a variety of perception and prediction problems. Deep learning has transformed vision systems that are sometimes better at recognizing objects in real data than people. It has revolutionized natural language processing tasks, such as the spoken word understanding system in our smartphones and leading audio analysis systems [3]. In the process, it is revolutionizing the consumer experience and simplifying many processes.

W. Shafik, *The Dark Side of AI*, https://doi.org/10.1007/978-3-032-09130-7_10

10.1.1 Definition and Scope

This white paper uses the term AI at a high level to refer to activities and systems that are designed, at least in part, to mimic, augment, or enhance human cognitive and decision-making capacities, as well as activities and systems adopted by humans that use some form of machine learning. The term AI covers disparate domains of technological innovation, including machine learning, computer vision, audio analytics, and natural language processing, as well as the wide scope of endpoints, such as recommendation systems, fraud detection, credit decision-making, digital assistants, and autonomous systems, to name a few. Many of these innovations are components of broader AI systems, where design, implementation, and use of these AI components frequently have unique and context-specific challenges. Therefore, the appropriate approach to securing these systems may differ, and each topic, from AI fairness to AI explainability, deserves due consideration [4].

Machine learning—particularly deep learning—has seen rapid progress in the past decade or so, leading to the development and application of a variety of novel and powerful AI tools to achieve progress in near-human or, at times, super-human performance in specific problem domains. Although the unique capabilities of these tools have made it possible to autonomously solve extremely challenging, high-impact problems that previously required human effort and judgment, their use to build reliable, scalable, and safe autonomous systems requires navigating a number of challenges: economic, operational, and ethical. These issues are difficult and often require the input of a diverse set of stakeholders, but are important to address to manage the deployment of AI systems effectively [5]. The functioning of these systems underpins many aspects of our modern economy, society, and personal lives, and they range from algorithmic detection of suspicious images on social media to autonomous vehicles and medical diagnostic systems that assist front-line workers in making critical decisions [6].

10.1.2 Historical Context

Many of the debates in AI reflect questions that have preoccupied philosophers and thinkers for centuries. In the seventeenth century, intelligence was defined in terms of the ability to reason and reflect, while the idea of non-human animals having consciousness was rejected in the early eighteenth century [7]. Later, in the nineteenth century, intelligence was embedded within the framework of an analytical engine, leading to the emergence of computer science theory. In any case, it is increasingly recognized by experts from different fields that most of the challenges raised by AI are not new, but rather, are simply taking on a new technological face. In fact, it also appears that concerns about artificial intelligence have grown exponentially in recent decades, possibly following the acceleration of the capabilities of these technologies [8]. The increase in interest in artificial intelligence can be attributed to a complex process in which multiple factors have interacted, such as the theoretical and practical developments of computer science, the institutional support for research and technological innovation, coupled with aggregate demand, and the resource constraints, especially on public policy. It can be hypothesized that certain historical events, by stimulating investments for civilian and military advances in computer science, have also contributed to the development of both the tools enabling big data operations and the first theoretical developments in artificial intelligence, helping to pave the way for what is referred to as the first age of digital technology. The exponentially growing interest in the theme was consolidated most recently, along with the exponential increase in available data and computers' computational power, after 2000 [9]. Indeed, investments followed, a new interest in the topic developed, and a new technological age has started to emerge, driven by the advancements in artificial intelligence and its application.

10.1.3 Current Trends in AI Technology

The intersection of present AI technologies with other powerful economic and social trends of our time is helping to drive a sense of urgency and excitement today, which is in turn accelerating AI research and deployment. Consider trends in some of these areas, as well as how they intersect with present AI capabilities. First, there are significant trends in how people and systems generate and interpret data [10]. The advent of the internet and the web has made it easy to publish and access vast amounts of structured and unstructured data. Many physical systems are now accessible and controllable remotely via the internet. On the privacy front, while there are significant social and policy debates around these trends, it is clear that people are voting through their actions for more digitization of intimate aspects of their lives and less privacy [11]. The operationalization of big data ideas, with concomitant improvements in data access, storage, processing, and analytics, has transformed business, government, health, and many other domains. Social scientists were early leaders in developing the analytics that could make sense of the new data swell, but with more researchers and business leaders looking for ways to understand and analyze increasingly complex and high-dimensional data, it wasn't long before these individuals turned to machine learning and statistical AI technologies to address their information processing needs. These technologies are now an important part of the analytics repertoire of social and business scientists, and their acceptance is likely to grow as future generations of data are available, also partly because of ongoing AI research [12].

10.2 Human-Centered Design Principles

In AI design and delivery, we must center the needs of people, including their capabilities, vulnerabilities, and limitations. To guide the development of AI systems that serve human needs and support human diversity, six principles of human-centered AI have been adopted. These are our first attempts to illuminate the human-centered ways in which we want our AI systems to be developed and operated. Of course, these principles will continue to evolve, with the field of AI as a whole, the technology landscape, and the broader world [13]. The six principles are designed to guide the development and use of AI innovation in a manner that earns people's trust and respect. They were informed by a steering committee of leading experts from a range of disciplines, including AI, machine learning, economics, social psychology, public policy, philosophy, design, law, and others. The principles cover the following areas: Fairness, Reliability and Safety, Privacy and Security, Inclusivity, Transparency, and Accountability. We hope that these guiding principles will enable us as a community to shape a technologically advanced future that embodies our collective humanity [14]. And that our principles will serve as a beacon for future progress that respects and supports the richness of personal identity, both past and present, as it continues to evolve in the future.

10.2.1 User Experience and Accessibility

In the project's research, user experience (UX) is defined as the human experience with different tools that support work, such as applications, devices, and services. The user experience influences work attitudes, productivity, data quality, and work satisfaction, and is a particular focus of public service research, where improving federal websites' UX is always a critical initiative [15]. Users frequently access federal websites to access services, register businesses, file taxes, access information, or complete tasks. Usability and accessibility are key components of the user experience, which aim to help people access information and accomplish tasks quicker and with less effort. When websites improve the magnitudes of those components, communication becomes more accessible, and task efficiency increases [16]. Crucially, the utility of government functions is enhanced for everyone, despite any disabilities or levels of digital proficiency. The study compared usability and accessibility issues and ratings of a selection of federal websites from two different customer experience evaluation organizations from the public and private sectors, using different combinations of web tools, guidelines, and best practices. Overall, the main hypotheses were supported. Although it is customary to think about usability and accessibility separately, the results showed a strong correlation between both measures. The more distinct websites, the fewer issues with usability and accessibility. Government websites were rated as less usable and less accessible than the rest of the websites, and government sites with fewer reported issues were found to have higher user experience scores [17]. Therefore, the number of utilities and best practices identified, as opposed to the tools and guidelines, was noted to influence the number of issues reported in the evaluations.

10.2.2 Inclusivity in AI Development

Responsible AI development must fundamentally address the source of the problems in how AI is created. Diverse and inclusive development teams will work to recognize and mitigate harmful applications of AI in advance. It is specific to this point in the process of AI development that inclusivity holds unique importance, essentially as a benignity checkpoint for AI innovations before they can be made widely available to society. Without diverse input, it's all too easy to develop AIs biased in favor of certain groups unknowingly. The first order fixes are practical and logistical. Bringing more people into the process should happen in parallel to a recognition of the unique values of diversity in AI design [18]. To achieve diverse and inclusive AI development communities, the AI education pipeline must be radically expanded in a way that is open to a wide, representative audience. Online tools are highly scalable and can reach a wide range of the global population. There are also existing initiatives bringing new voices into these debates. These organizations should be empowered to tackle these new challenges to open discussion and inclusive representation head-on. This process, or similar ones, could be a way to introduce new representative voices into the AI regulatory process [19]. The people determining AI policy must include a diverse group of stakeholders who can bring different perspectives to the issue.

10.3 Regulatory Frameworks

As AI technologies evolve and are increasingly deployed in mission-critical or safety–critical roles, there will likely be growing calls for the regulation of AI. Some believe that a proactive and overarching regulatory framework for AI can prevent both the real and imagined risks of AI today [20]. However, the views on the risks of AI diverge widely. AI is increasingly relied upon to make mission-critical and safety–critical decisions. Even if AI-based decision-making remains below human levels of reliability, the understanding of and capabilities to construct certification and compliance requirements with respect to degrees of AI perfection depend on the regulatory theory developed across industry sectors and countries [21]. At present, this understanding is at a nascent stage, although several societies and expert communities are enthusiastic about developing an explosion of legal scholarship on AI. One possible regulatory framework for AI consists of combinations of private safeguards, public regulatory standards, court-supported sanctions, and flexible verification and auditing procedures to correct unexpected AI decisions. Acceptance criteria for AI systems must be connected to the organizational processes used to generate and validate the systems. In the modern world, AI may be taken to be a proxy for intelligent agents such as institutions, organizations, or humans, or even combinations of these within mixed human-AI groupings, who are felt to regulate themselves or be regulated in ways that differ from those who remain outside AI systems [22]. Proper governance design arrangements could solve or at least limit challenges and issues related to the use of AI, especially with respect to the principles of data control and privacy, transparency, and accountability.

10.3.1 Global Standards and Guidelines

AI policy, ethics, and safety governance considerations are essential in designing and building AI systems. Governments around the world should consider and address these considerations in their operating plans and competitive sourcing process development. Global standards and guidelines for AI should be established to help drive trust and effort toward more ethical and safety-enhancing AI design and deployment. Only by having a united approach to AI systems design and development will we be able to realize the incredible advancements that AI has to offer more quickly [23]. There is a recognized need for the establishment of global sets of 'AI by Design Principles' that will emphasize a need for 'human-centric' values. Establishment may require a known enforcement trend, but the economic and geopolitical impact of such design principles would potentially set the framework of future global norms for AI development. AI by Design is a vision to counter many of the negative implications before AI systems are created. This is beneficial both economically and socially, creating frameworks based on the ethical concepts most people have today [24]. Ensuring a path of economic growth from the invention and maintenance of such an AI-empowered society, ensuring that society uses it.

10.3.2 National Policies and Regulations

National policymakers are responsible for ensuring that AI applications are developed, used, and governed in a way that respects human rights, protects the public interest, and contributes to the overall well-being of society. They should establish and reinforce mechanisms of oversight, conduct regular audits of the impacts of AI systems, and ensure the existence of transparent and effective accountability and redress procedures that address the risks created or exacerbated by AI. Government institutions should lead by example and engage in the responsible adoption of AI while also promoting its development through public–private collaboration. Last but not least, policymakers should earmark adequate resources for research and education and create incentives to foster an ethical and human-centric approach to AI across all sectors [25]. Collectively, these individual policy choices will help design a global ecosystem that promotes the kind of AI we need in order to live better and fuller lives.

Given the significant transformative potential of AI, it seems natural that national governments will want to establish policy frameworks that ensure that the benefits exceed the risks. Making informed policy decisions requires at least a basic understanding of what AI is and what it enables and entails. AI has the potential to evolve rapidly and to affect a wide range of sectors. Regulatory frameworks will need to adapt [26]. They should be closely coordinated at both national and international levels and should be developed and tested iteratively and with broad stakeholder input, with initial rules being developed to address the most significant concerns, such as safety, security, privacy, and transparency. As the use of the technology deepens, the role of regulation should evolve from one of foundational protection—ensuring that AI technologies operate in a manner consistent with societal norms—to a more interventionist approach—either promoting or discouraging certain types of behavior or particular applications of AI based on a deeper understanding of the impact on society [27]. All this should benefit from public sector expertise, stakeholder input, and transparent decision-making. Promoting knowledge exchange among nations—including capacity-building elsewhere in the world—should also be a priority.

National policymakers play a key role in promoting a well-being-driven approach to the design and use of AI through the development and enforcement of a holistic framework for responsible AI. Such a regulatory framework should be designed with flexibility and scope in mind in order to ensure that policy choices can keep up with the rapidly evolving and diffuse nature of AI, its vertical and horizontal applications, and its impacts on individuals, societies, and governments. Regulatory options should be chosen to support the AI ecosystem at the intersection of research, industrial, financial, environmental, and social policies [28]. Regulations, directives, and soft instruments should embody and foster principles of human rights, accountability, transparency, non-discrimination, fairness, and user-centricity. At the same time, they should incentivize research and innovation while leaving space for voluntary initiatives. In addition, both regulatory bodies and the private sector should invest in the definition of technical standards, allowing the implementation of the design principles everywhere in the world. In particular, all stakeholders are responsible for prototyping and demonstrating convenient and ready-to-use tools, which could make design principles actionable, thus showing their value; formalizing AI assessment and certification techniques as a complement to transparency and explainability; creating tests and benchmarks to make AI development teams focus on specific accountability dimensions, thus educating an AI-literate society; drafting the legal and normative implications of the above activities [29, 30].

10.4 Stakeholder Engagement

AI systems can embody unique technological risks and vulnerabilities and implicate important values such as non-discrimination, fairness, privacy, safety, transparency, explainability, reputation, and accountability. Together, fostering AI innovation while mitigating risk and balancing stakeholder interests is challenging, requiring input from a wide range of affected stakeholders. Persuading deep and sustainable stakeholder engagement regarding the challenges and policy inquiries I have highlighted above will also require the establishment of broad-based participatory processes to help ground the policymaking in democratic wisdom [31]. The AI policy domain should be characterized by strong, inclusive, and fair stakeholder engagement. Despite widespread consensus on the importance of public deliberation and diverse stakeholder input and consultation, there are reasons to worry that we are not living up to our commitment to pluralistic and participatory decision-making in the establishment of AI policy [32]. Right now, the AI policy conversation is unmoored from public participation, and institutionalized structures for participation are both conspicuous by their absence and important for our values. This absence is especially striking in light of the wide-ranging implications AI may have for a range of societal domains, including our public and community institutions, education, welfare, and employment systems, and individual human rights and responsibilities [18].

10.4.1 Role of Government

The U.S. government must play leading roles in supporting fundamental research, creating large-scale infrastructure and assembling data, enabling full participation and widespread access, addressing fundamental issues related to transparency, privacy, cybersecurity, and safety, modeling ethical thinking around novel AI implementations, and creating a stable environment that is friendly to public and private investment in research, development, and deployment of AI. As a funder of research and provider of infrastructure, the mission of the U.S. government is to ensure that the necessary foundational AI capabilities are appropriately hosted and robustly fostered to secure long-term national competitiveness and global technological leadership in AI invention that is fundamental to advancing the U.S. mission in economics, national defense, and secure governance and human character [33]. The U.S. government should explore new ways to support the broadening of AI education, including lifetime learning, defining new curricula in related fields, creating opportunities for practical experience, and offering applied fellowships and internships for engineering and science students. Government agencies should serve as many of the AI project hosts as part of their operational mission, that are ready for broader use, provide researchers and entrepreneurs with unprecedented access, and use standardized data collection to develop AI solutions with societally beneficial impacts [34]. As AI-dependent services and programs are central, government agencies should provide support for citizens who need them most, to ensure that members of these groups are aware of the presence and availability, hence accessibility, of government-supported AI solutions.

10.4.2 Industry Responsibilities

The launch and implementation of the AI Principles could have a significant positive effect on global efforts to create AI solutions that are ethical, transparent, and fair. Businesses need to devise ethical digital AI strategies whereby they consider the potential harms or risks posed by their AI systems or AI-based outputs to people and societies. These ethical digital strategies need to be comprehensive and go beyond mere compliance with legislation. Clear and enforceable principles should govern the AI ecosystem in which businesses coexist and operate, where businesses are required to use AI ethically and consciously, and external factors, like the purpose for which AI is being used and the possible impact on jobs and wages, are considered [35]. On its journey towards AI resiliency, business stakeholders should involve a variety of top management and gain industry knowledge through courses, advocacy work, and academic partnerships. Furthermore, businesses are responsible for making a variety of internal and external investments in order to meet their ethical and conscious digital AI use responsibilities. They also have a role to provide better, ever more productive instruments and instruction for stakeholders outside of the business domain. In order for businesses to operate in an increasingly ethical manner, more guidance, insightful case studies, and collaboration are required to link the understanding of AI in different sectors and to promote coherence. Co-creation of values based on global ethical principles must guide the ongoing emergence of AI, including multi-stakeholder discussions and agreements [36].

10.4.3 Community Involvement

We assume that AI should be firmly rooted in and guided by the needs, goals, dreams, and hopes of people. We want to avoid what has transpired in social networking and e-commerce platforms. These are under constant public scrutiny because they seem to treat the users as mere strings of data, data that can be manipulated. We must not allow AI to follow in this tradition. These are the principles that should guide the development of AI. But this will be effective and can only be achieved if the members of the community are unanimous and steadfast in their support. This is why it is so important to have open, unrestricted community involvement in deciding the future course of AI development [37]. Everyone, especially those who are going to bear the brunt of the operations, must be part of the decision-making process. Community involvement should reach everyone, from business and civic leaders to community organizers, and, more importantly, to local populations throughout the world. The obvious question is: How can we implement community involvement? We, in the AI research, business, and development community, must develop a stakeholders' group that has a widely inclusive and influential voice. We suggest that the group be called the Initiative for Community Engagement and Progress. This is a hard question, but it is a crucial one. Even if it is hard, we must meet the challenge and succeed [38]. The advancement of AI and the benefits it brings to everyone rest upon such success, as presented in Table 10.1.

10.5 Case Studies of Responsible AI

Companies, intergovernmental organizations, and civil society have emphasized the importance of AI for humanitarian action, but many valuable examples of AI for good go unrecognized. There are a large number of case studies beyond those profiled here, particularly among smaller firms that do not have visibility in the broader public debate [39]. This section presents case studies from development, health, disability, justice, and media and focuses on applications that advance human capabilities, foster better public decision-making, and are aligned with the UN's sustainable development goals. These examples illustrate how AI can be used to improve people's lives in ways that make sense for societal well-being and not just economic growth [40]. They show how AI can increase a sense of community for marginalized or underprivileged people. One prevalent example of machine learning and AI being used to combat discrimination is the introduction of randomized admissions decisions in the Korean college and university system. Korean students' parents are well known for their emphasis on education and the sacrifices that they are willing to make to help their children succeed. Most adults in Korea can tell stories of students who are the first in their families to have an opportunity to attend college, but they can also share many stories reflecting a more unfortunate situation: admissions are controlled primarily via a student's marks, and many students do not perform well on a narrow range of measures because, as they get older, they feel obliged to work part-time jobs to help their families [41]. AI, in the form of machine learning, is being deployed to help design randomized admissions decisions in Korea so that students from less advantaged families and students working long hours to make ends meet will have a fair opportunity to qualify and have an increased chance of being the first student or one of the few students from their schools and communities to have the possibility to attend college with students of privilege. These types of AI projects should be encouraged because they challenge stereotypes, given the potential to obtain positive societal outcomes with such a trivial amount of total effort from those associated with privilege [42].

Table 10.1 Building blocks of a human-centered and responsible AI ecosystem

Key pillar	Goal	Strategy	Stakeholders involved	Tools/frameworks	Outcome
Human-centered design	Prioritize human needs and values	Co-design with users and communities	Designers, end-users	Human-centered AI toolkit	Greater usability and societal relevance
Ethical AI development	Prevent harm and uphold rights	Ethical impact assessments, bias audits	Developers, ethicists	IEEE ethically aligned design	Safer and fairer AI systems
Transparency and explainability	Enhance understanding and trust	Deploy explainable AI models	AI engineers, regulators	SHAP, LIME, AI explainability 360	Informed users and greater system trust
Inclusivity and diversity	Avoid marginalization and bias	Engage diverse teams and data sources	HR, data scientists	Inclusive design principles	Equitable access and representation
Accountability and oversight	Ensure responsibility for AI actions	AI audit trails, governance boards	Organizations, governments	Algorithmic accountability frameworks	Ethical and legal compliance
Data privacy and security	Protect personal and sensitive data	Use of privacy-preserving techniques	Data custodians, legal advisors	Federated learning, differential privacy	Respect for user autonomy and rights
AI education and literacy	Promote understanding and critical use	Public awareness campaigns, curricula	Educators, NGOs	AI4K12, elements of AI	Empowered and informed citizens
Environmental sustainability	Reduce AI's ecological footprint	Optimize algorithms, green infrastructure	AI researchers, cloud providers	Green AI practices, carbon-aware computing	Sustainable tech ecosystems
Policy and regulation	Guide safe deployment and innovation	Develop ethical laws and global norms	Policymakers, international bodies	EU AI act, OECD AI principles	Balanced innovation and public protection
Multistakeholder collaboration	Create inclusive and robust solutions	Build international, cross-sector alliances	Academia, industry, civil society	Partnership on AI, GPAI	Shared knowledge and global alignment

10.5.1 Successful Implementations

In a recent survey, we asked companies about the factors that are critical to successful AI implementation. For many companies, achieving positive ROI is a key measure of success. The chart below provides additional insight. Companies behind the most successful projects are more focused on strategic wins, such as increased revenue, faster innovation, and broader market applications. They also tend to use more types of AI technologies, set ambitious goals, and use measurement methods. Successful AI projects also receive more investment from companies. For those companies that have had the best experiences with AI, their approach is markedly different, and they are betting bigger. They use more AI technologies, set more ambitious targets, and demand more from employees to meet and exceed goals. From the data, it is clear that companies that expect the most extreme AI outcomes also have the most AI projects that are indeed artificially intelligent [43]. They employ convolutional neural networks, generative adversarial networks, and decision trees, among other technologies, more frequently in their AI work than the more conservative companies.

10.5.2 Lessons Learned from Failures

AI research and practice carry their kinds of lessons about how to behave better in the future by looking at failures and breakdowns. There are both specific and general lessons to be learned from the experience of AI breakdowns. Some, such as the knowledge that systems will act on the training data, are continually relearned and forgotten. Others are completely new insights, resulting from failures of systems that use techniques like machine learning for clear application areas such as image or speech recognition. We consider what AI research and practice can teach us about improving design in the future [44]. The outside story contrasted with many researchers' core belief that learning systems work because they only say what they have seen before. Using human insights, including a sense of style, enabled software that was otherwise unreliable in its understanding of the test dataset to be recognized as dangerous. As the system was experimental, we found out why it was wrong easily. Systems being deployed in practice may already be behaving like this, but without humans to recognize the problem. As we develop machine learning techniques to understand the motivations of systems that are not clearly explainable to humans, and thus only act on things they have been taught, we must frequently be reminded of this risk [45].

10.6 Future Directions for AI

There is a clear opportunity to apply our recommendations for the responsible development and use of AI to other emerging technologies. As AI matures, it is important to continue to engage with stakeholders in order to adjust the approach to risk mitigation. The task of charting a path ahead for the nation will require sustained engagement with academia, industry, nonprofits, and the public. More research will also be necessary to assist with the future design of AI systems that incorporate ethical considerations [46]. Cross-disciplinary research is necessary in order to develop AI systems that incorporate a wide range of values and respect people's rights and dignity. While technical progress is important, it is necessary but not sufficient to ensure the responsible development of AI. More research is needed to help develop international standards that contribute to the responsible development and use of AI. All our recommendations suggested that the U.S. should lead in serving as a model in this area, which would be bolstered by having a national strategy on AI development. Without a government monolith guiding a national strategy, the United States should continue to engage in the national and AI arena [38]. What would assist in this effort is to continue to work with like-minded partners to ensure a future in which AI respects human rights, freedom, and safety.

10.6.1 Emerging Technologies

Emerging technologies hold tremendous promise, from addressing problems in medicine and agriculture to promoting peaceful, prosperous, and inclusive societies. However, each new advance raises complex ethical and societal issues. It's not enough to adhere to the principle of "not harm"; we need to promote broad societal benefits actively. Disparities in global resources and social rules will affect the pace and shape of technological advancement. We seek input from stakeholders around the globe to ensure that technology thrives everywhere. Computerized artificial understanding, which is on the verge of science fiction today, could catalyze a new industrial revolution far beyond anything we've seen before [29]. If we fail to develop AI safety mechanisms, do not provide stewardship for the future stage of human evolution, or leave massive segments of our population without jobs or dignity, we could easily lose the race to the future, having instead moved to a future where machines work for other machines and demand very little that the Earth could support. The challenge to modern society is very clear: it is our task to guide this relationship in ways that are beneficial to current and future generations, rather than tailing it. All people worldwide need to ask themselves the question, "What sort of society do we hope to create for our great-grandchildren?" and then take individual and collective paths to get there [28]. Truly, this is one of the twenty-first century's largest worthwhile challenges.

10.6.2 Long-Term Impacts on Society

In the long term, AI will impact the way people relate to each other, as well as the overall evolution of humankind. From a technological evolution perspective, a well-known point of friction is that humans need to drive their technological improvement for AI to benefit humankind, but doing so would eventually push humanity out of the "driver's seat" and into an AI-driven society. Ethical considerations and understanding of the complexity of our nature can increase society's resilience against Deceptively Intelligent AI scenarios and accidental AI risks [5]. A more nuanced and empathic understanding of the complexity of human environments, governance, and perception can guide us toward the design of AI systems that are "friendly," ethically benevolent, aligned with human values, and focused on ensuring the well-being of future generations as well. The advancement and use of AI technologies on their own are not inherently good or bad; it depends on the principles by which these tools are used. Humankind's opportunity is to develop AI that supports human actions and decisions, with awareness and charity, in order to achieve the common good. Although standard AI research often focuses on creating AI to be fast, intelligent, and extended, ethical AI research should focus on pouring its best potential intelligence into moral values. Our challenge is to develop AI that is "not too smart in solving secondary goals." The creation of this responsible AI is deeply linked to the recognition that AI systems are an important and interconnected part of the human community and environment [9, 12, 14]. Given the direct relationship between human flourishing and meaningful AI research, our goal is clear: to work to inspire AI researchers who want to contribute to the common good through their technical work.

10.7 Conclusion

The insights we have gathered from a wide spectrum of experts, researchers, and practitioners provide a comprehensive account of the global conversation around how to create AI that benefits society. We hope that this book invites even broader communities to join the conversation and equips ongoing initiatives with an up-to-date picture of the key opportunities and challenges in realizing a positive future for AI. Technologies such as AI that affect all members of society need to be developed in a human-centered way to ensure diverse perspectives are included in the process. Collaborations across disciplines such as computer science, law, public policy, the social sciences, and the humanities are essential to encourage, nurture, and ultimately prioritize responsible thinking and practices in the AI ecosystem. Important next steps to foster a responsible AI research and policy agenda include discussions around global mechanisms and initiatives to facilitate capacities for responsible AI and more writings to complement the available draft principles by exploring the governance implications of living up to them. All stakeholders in the AI ecosystem, including governments, intergovernmental organizations, civil society institutions, investors, AI developers, insurers, citizens, and those who stand to profit from AI deployment, should ensure AI is developed and used with consideration of moral and ethical consequences, a respect for human rights, and respect for the environment. These stakeholders must be determined to work together to design and implement concrete strategies to foster the ethos of we-power, to strengthen the will to muster the intellect to create and ensure a responsible AI sector. The aim is to create a collective environment to shape an AI that will be able to create, with the help of the contributions of all, a society where all stakeholders can express the best in being human. It will be constant education and awareness programs that will enable all partners in the AI ecosystem to know the technology and responsibility that STEM includes in their design. We will know that we are moving towards this society when exploring further, as a positive signal for what already exists: investments in research and policy that will scale and operationalize responsible AI, data science, and new ICT systems.

References

1. Shafik W. Machine learning for advanced wireless communication: applications, challenges, problems, and open research questions. In: Microwave devices and circuits for advanced wireless communication. CRC Press; 2024. p. 252–78.
2. Holzinger A, Saranti A, Angerschmid A, Retzlaff CO, Gronauer A, Pejakovic V et al. Needs human-centered AI : challenges and future directions. Sensors. 2022;22.
3. Dong M, Bonnefon JF, Rahwan I. Toward human-centered AI management: methodological challenges and future directions. Technovation. 2024;131.
4. Denno P. Cognitive work in future manufacturing systems: human-centered AI for joint work with models. J Integr Des Process Sci. 2024;27.
5. Wang L, Zhang Z, Wang D, Cao W, Zhou X, Zhang P et al. Human-centered design and evaluation of AI-empowered clinical decision support systems: a systematic review. Front Comput Sci. 2023;5.
6. Shafik W. The role of generative artificial intelligence in e-commerce fraud detection and prevention. In: Strategies for E-commerce data security: cloud, blockchain, AI, and machine learning. IGI Global; 2024. p. 430–69.
7. Shafik W. Introduction to ChatGPT. In: Advanced applications of generative AI and natural language processing models. 2023.
8. Xu W, Dainoff MJ, Ge L, Gao Z. From human-computer interaction to human-AI interaction: new challenges and opportunities for enabling human-centered AI. Int J Hum Comput Interact. 2021;39(3).
9. Komasawa N, Yokohira M. Learner-centered experience-based medical education in an AI-driven society: a literature review. Cureus. 2023.
10. Shafik W. Agentic AI in personalized medicine and treatment: healthcare redefined, really? In: Integrating AI with haptic systems for smarter healthcare solutions. IGI Global Scientific Publishing; 2025. p. 85–100.
11. Carriço G. The EU and artificial intelligence: a human-centred perspective. Euro View. 2018;17(1).
12. Xu W, Gao Z, Ge L. New research paradigms and agenda of human factors science in the intelligence era. Acta Psychologica Sinica. 2024;56(3).

13. Bond RR, Mulvenna M, Wang H. Human centered artificial intelligence: weaving UX into algorithmic decision making. In: RoCHI 2019: international conference on human-computer interaction. 2019.
14. Bühler MM, Jelinek T, Nübel K. Training and preparing tomorrow's workforce for the fourth industrial revolution. Educ Sci. 2022;12.
15. Shafik W. Human-computer interaction (HCI) technologies in socially-enabled artificial intelligence. In: Future of digital technology and AI in social sectors. IGI Global; 2025. p. 121–50.
16. Rony MKK, Kayesh I, Bala S Das, Akter F, Parvin MR. Artificial intelligence in future nursing care: exploring perspectives of nursing professionals—A descriptive qualitative study. Heliyon. 2024;10(4).
17. Rowe JP, Lester JC. Artificial intelligence for personalized preventive adolescent healthcare. J Adolesc Health. 2020.
18. Kaluarachchi T, Reis A, Nanayakkara S. A review of recent deep learning approaches in human-centered machine learning. Sensors. 2021;21.
19. Rezaev A V., Tregubova ND. ChatGPT and AI in the universities: an introduction to the near future. Vysshee Obrazovanie v Rossii. 2023;32(6).
20. Shafik W. Generative adversarial networks: security, privacy, and ethical considerations. In: Generative artificial intelligence (AI) approaches for industrial applications. Springer; 2025. p. 93–117.
21. Visram S, Leyden D, Annesley O, Bappa D, Sebire NJ. Engaging children and young people on the potential role of artificial intelligence in medicine. Pediatr Res. 2023;93(2).
22. Schoenherr JR, Abbas R, Michael K, Rivas P, Anderson TD. Designing AI using a human-centered approach: explainability and accuracy toward trustworthiness. IEEE Trans Technol Soc. 2023;4(1).
23. Lawrence LEM, Echeverria V, Yang K, Aleven V, Rummel N. How teachers conceptualise shared control with an AI co-orchestration tool: a multiyear teacher-centred design process. Br J Educ Technol. 2024;55(3).
24. Lin S. A clinician's guide to artificial intelligence (AI): why and how primary care should lead the health care AI revolution. J Am Board Family Med. 2022;35(1).
25. Lu C, Lyu J, Zhang L, Gong A, Fan Y, Yan J, et al. Nuclear power plants with artificial intelligence in industry 4.0 era: top-level design and current applications—A systemic review. IEEE Access. 2020;8.
26. Shafik W. Toward a more ethical future of artificial intelligence and data science. In: The ethical frontier of AI and data analysis. IGI Global; 2024. p. 362–88. https://services.igi-global.com/resolvedoi/resolve.aspx?doi=10.4018/979-8-3693-2964-1.ch022.
27. Khosravy M, Gupta N, Pasquali A, Dey N, Crespo RG, Witkowski O. Human-collaborative artificial intelligence along with social values in industry 5.0: a survey of the state-of-the-art. IEEE Trans Cogn Dev Syst. 2024;16(1).
28. Wienrich C, Latoschik ME. eXtended artificial intelligence: new prospects of human-AI interaction research. Front Virtual Real. 2021;2.
29. Karpus J, Krüger A, Verba JT, Bahrami B, Deroy O. Algorithm exploitation: humans are keen to exploit benevolent AI. iScience. 2021;24(6).
30. Shafik W. Deep learning impacts in the field of artificial intelligence. In: Deep learning concepts in operations research. New York: Auerbach Publications; 2024. p. 9–26. https://www.taylorfrancis.com/books/9781003433309/chapters/doi.org/10.1201/9781003433309-2.
31. Shafik W, Kalinaki K, Fahim KE, Adam M. Safeguarding data privacy and security in federated learning systems. In: Federated deep learning for healthcare. Boca Raton: CRC Press; 2024. p. 170–90. https://www.taylorfrancis.com/books/9781032694870/chapters/doi.org/10.1201/9781032694870-13
32. Shafik W. Science of emotional intelligence. In: Enhancing and predicting digital consumer behavior with AI. IGI Global; 2024. p. 284–310. https://services.igi-global.com/resolvedoi/resolve.aspx?doi=10.4018/979-8-3693-4453-8.ch015.

33. Scott Hansen S. Public AI imaginaries: how the debate on artificial intelligence was covered in Danish newspapers and magazines 1956–2021. Nordicom Rev. 2022;43.
34. Ahmad K, Maabreh M, Ghaly M, Khan K, Qadir J, Al-Fuqaha A. Developing future human-centered smart cities: critical analysis of smart city security, data management, and ethical challenges. Comput Sci Rev. 2022;43.
35. Capel T, Brereton M. What is human-centered about human-centered AI? A map of the research landscape. In: Conference on human factors in computing systems—Proceedings. 2023.
36. El-Assady M, Moruzzi C. Which Biases and reasoning pitfalls do explanations trigger decomposing communication processes in human-AI interaction. IEEE Comput Graph Appl. 2022;42(6).
37. Gu H, Huang J, Hung L, Chen XA. Lessons learned from designing an AI-enabled diagnosis tool for pathologists. Proc ACM Hum Comput Interact. 2021;5(CSCW1).
38. Liao QV, Wortman Vaughan J. AI transparency in the age of LLMs: a human-centered research roadmap. Harv Data Sci Rev. 2024;(Special Issue 5).
39. Wasswa S. The role of emotional intelligence in career development and career success. In: Pushan KD, Sachin G, Shafali K, Anita G, Rita Karmakar, Pronaya Bhattacharya, editors. Emotional intelligence in the digital era: concepts, frameworks, and applications, 1st ed. Taylor & Francis Group; 2025. p. 108–28.
40. Morrison K, Shin D, Holstein K, Perer A. Evaluating the impact of human explanation strategies on human-AI visual decision-making. Proc ACM Hum Comput Interact. 2023;7(CSCW1).
41. Lehtiö A, Hartikainen M, Ala-Luopa S, Olsson T, Väänänen K. Understanding citizen perceptions of AI in the smart city. AI Soc. 2023;38(3).
42. Thieme A, Hanratty M, Lyons M, Palacios J, Marques RF, Morrison C, et al. Designing human-centered AI for mental health: developing clinically relevant applications for online CBT treatment. ACM Trans Comput-Hum Interact. 2023;30(2).
43. Asan O, Choudhury A. Research trends in artificial intelligence applications in human factors health care: mapping review. JMIR Hum Factors. 2021;8.
44. Berretta S, Tausch A, Ontrup G, Gilles B, Peifer C, Kluge A. Defining human-AI teaming the human-centered way: a scoping review and network analysis. Front Artif Intell. 2023;6.
45. Lee S, Lee M, Lee S. What if artificial intelligence become completely ambient in our daily lives? Exploring future human-AI interaction through high fidelity illustrations. Int J Hum Comput Interact. 2023;39(7).
46. Borsci S, Lehtola VV, Nex F, Yang MY, Augustijn EW, Bagheriye L, et al. Embedding artificial intelligence in society: looking beyond the EU AI master plan using the culture cycle. AI Soc. 2023;38(4).